CHRIS CAGE

Still Human

Staying Sane, Productive, and Fully You in the Age of AI

THE
MENTAL
LENS

First edition

ISBN: 979-8-9931606-1-0

This book was professionally typeset on Reedsy.
Find out more at reedsy.com

For Laura and the boys. Thank you for keeping me human when the world tries to make me otherwise.

"The machine does not isolate man from the great problems of nature but plunges him more deeply into them."

Antoine de Saint-Exupéry

Contents

Preface

A Note to the Reader

This book is not a scholarly research publication. It is written in a personal voice, blending my own experiences with insights from psychology, neuroscience, and technology. I reference studies and research throughout to support the ideas, but my goal is not to provide an exhaustive academic review. Instead, I aim to share stories, tools, and reflections that make mental health and productivity more relatable in the age of AI.

If you are seeking peer-reviewed articles, you will find citations in the Research Notes and Sources section at the end. But above all, I hope you find this book approachable, human, and valuable on your own journey.

Introduction

Welcome to the Machine (and Why You Don't Have to Become One)

We have more tools than ever; apps that plan our day, watches that track our sleep, and assistants that finish our sentences. So why does life feel more complicated to manage? Why do we feel more behind, more anxious, more disconnected, even as technology promises that we will finally be in control?

One Saturday morning, while I was making pancakes, my two kids started fighting over the smart speaker. One wanted Alexa to play the "animal of the day" video, and the other insisted on listening to some gummy bear song so he could dance. Meanwhile, my phone buzzed with a reminder to take my medicine and my watch tapped me on the wrist to get a stand goal...While I was already standing.

I love technology. I have the wireless mesh network, voice assistants in every room, productivity apps, Artificial Intelligence (AI) research tools, and smart everything. I did not get into all this tech because I was overwhelmed. I got into it because I was excited. I wanted the future; the frictionless, optimized, push-button life. And in many ways, I got it. But I did not expect the cognitive clutter that came with it. My house

1

is so connected it practically hums. My brain, on the other hand? Less humming. More... buzzing. Loudly.

It started innocently enough. I picked up an Amazon Echo to help play music in the kitchen. Great! Then one for the bedroom. Then the kids' room. Then the bathroom, because why shouldn't you be able to check the weather while brushing your teeth? Before I knew it, I had a network of Alexa's chirping away in nearly every room of the house, all delivering half-helpful answers to hesitant questions, mostly asked by my kids. One of them once asked Alexa which shark was the strongest, and it triggered an existential crisis. Not for the kid, for me.

Add in my iPhone, iPad, personal laptop, work laptop, and half a dozen apps all trying to send me "urgent" reminders, and it is not just a smart home anymore. It is a cognitive circus. A digital carnival where you are the ringmaster, the clown, and the slightly unhinged lion trying to escape.

This is not just information overload. It is decision fatigue, dopamine spikes that trigger the feel-good reward signal in your brain, guilt about tasks left unfinished, and anxiety over the ones you completed. It's like running a race where the finish line keeps moving every time you get close.

Most of these tools are genuinely helpful. The wireless mesh network? Lifesaver. OpenAI's ChatGPT? It has changed how I research, write, and run my blog and podcast. I built The Mental Lens website in less than two weeks, thanks to AI; something that would have taken me months otherwise.

But even with all this tech, or maybe because of it, I often ask myself:

why is everything still so hard to manage?

The overwhelm is not just about information or noise; it is about the mental toll of trying to stay on top of it all. The apps that promise to organize our lives end up taking over them. The tools meant to free our time end up filling it with notifications. The tech designed to make life easier slowly rewires how we think, connect, and rest.

And then there is the guilt, that subtle, constant pressure that we should be more efficient, more optimized, more on top of things. Even when we are doing "all the right things", it still feels like we are behind.

I used to manage my entire to-do list in a spreadsheet. It worked for years with color-coded tabs, manually updated priorities, and the whole system. But eventually, it became a burden. Outdated. Slow. Manual. Like trying to run a modern operating system on a flip phone.

We live in a world that is fast-moving, always-on, and AI-accelerated. And yet... we are still human. Still tired. Still trying to feel like we are in control.

This book is for anyone who feels caught in that tension, the one between loving what tech makes possible and hating how it makes them feel. Between curiosity and burnout. Between wanting to be productive and just wanting to feel okay.

We will explore this in three parts:

Part I: Why being "always on" is exhausting our brains, fraying our focus, and feeding anxiety.

Part II: How to redesign the way we work and think so that it works with our biology instead of against it.

Part III: How to protect your creativity, emotional depth, and humanity in a world that runs on code.

We are not here to reject AI or glorify the past. We are here to ask better questions:

How do we work with technology instead of drowning in it?

How do we use AI to support our mental health, not sabotage it?

And how do we build lives that feel grounded, calm, and meaningful - even as the world gets faster and smarter?

This is not about keeping up with the machines. It is about remembering what they cannot do for us: feel, imagine, care, dream. Those emotions are not inefficiencies. They are the feature, not the flaw.

And being human is not a flaw to fix.

It is the very essence of why we're here.

I

Part I: The Anxiety of Being "Always On"

"Men have become the tools of their tools." — *Henry David Thoreau*

1

Your Brain Was Not Built for This

Notifications, inboxes, endless browser tabs... Welcome to the modern panic room.

Let's start with a quick thought experiment: imagine you are a hunter-gatherer, 200,000 years ago. You wake up. You forage. You might hunt a squirrel, avoid a saber-toothed tiger, make eye contact with two humans all day, and then go to sleep when the sun goes down. Simple life. High stakes, sure. But simple.

Now compare that to this morning: You wake up to five notifications before your feet hit the floor. Your smartwatch buzzes. Your work's enterprise instant messaging solution is already pinging. You skim 37 new emails, forget to reply to the one that actually matters, toggle between calendar apps, check the weather, then open four browser tabs to catch up on the latest trends, none of which you actually read. You glance at a productivity app, update your to-do list, remember that you are out of coffee, and debate whether you have time to microwave your brain with one more podcast before your next meeting. It is 8:17 a.m. and your cortisol, your body's built-in stress hormone, is already clocked in.

Welcome to modern life, where your brain, a beautifully evolved machine designed for scanning threats, navigating social dynamics, and solving real-world problems, is now trying to survive group chats, Google Docs, auto-scheduling AI, digital assistants, and TikTok recipes all at once.

Your Ancient Brain, Upgraded Poorly

This is not just a "busy schedule" problem; it is an evolutionary mismatch.

Your brain's core survival systems have not had a proper software update in about 50,000 years. It is still wired to detect threats such as lions, tigers, or bears and to notice rewards like ripe fruit or social approval. Now, instead of predators and campfire gossip, we have notifications, algorithmic timelines, and pop-ups. And your brain treats every notification as a potential signal worth reacting to.

It is running on what psychologists call the "Stone Age Operating System" [1], a metaphor for evolutionary mismatch, where ancient survival wiring meets endless modern stimuli. Psychologists describe this mismatch as the consequence of having neural systems evolved under ancestral conditions now struggling to adapt to 21st-century life. The rational part of your brain, prefrontal cortex, is constantly battling your limbic system, the emotional and reactive part, while both are getting ambushed daily by dopamine hits from likes, comments, calendar reminders, and whatever AI-generated nonsense your inbox coughed up overnight.

That evolutionary mismatch: between how our brains evolved and the world we live in now, is one of the biggest drivers of chronic mental fatigue.

We were designed to focus intensely on one thing at a time, with generous pauses for rest. But now we expect ourselves to answer emails mid-conversation, solve work problems in the middle of dinner, and use AI to brainstorm, outline, edit, and schedule content before our coffee even kicks in.

It is not that we are lazy or distracted. It is that we are overloaded and constantly being pulled in competing directions by tools that were designed to make us more efficient, but which ultimately end up demanding more from us instead. The mismatch between how our brains evolved and the world we live in now is one of the biggest drivers of chronic mental fatigue.

Caveman Brain (200,000 BC)	Modern Brain (2025 AD)
Focused on one task: find food	Juggles 20 browser tabs before lunch
Social circle: 5-10 people	Chat messages, video calls, and DMs from strangers
Sleep with the sun	Blue light until 1 a.m., then doomscrolling
Run from tigers occasionally	Get 147 dopamine hits before breakfast
Natural silence and downtime	Constant digital noise and multitasking
Communicates face-to-face	Dictates to AI while texting during a meeting
Reward = fresh berries	Reward: likes, badges, and email zero

Caveman Brain versus Modern Brain

Your Brain's Sticky Note Is Already Full

Here is a lesser-known fact: your working memory, the mental scratch pad where you juggle thoughts, is tiny [2]. And not in a small-brain kind of way; everyone's working memory comes fun-sized.

Basically, it is the brain equivalent of a Post-it note, not a 10-terabyte cloud server. When you overload it with incoming information, something falls off before you have had a chance to process it.

In a simpler world, you would use that sticky note to track one or two important things, like where the food is or which direction the predator went.

Today, it is holding your grocery list, your next link for a video call, that random password you just created, and the fact that you owe your boss an email that just fell off the note because another work chat notification came in.

When Technology Took Over the House

I understand this feeling intimately.

I have always been passionate about technology, not in a "build my own computer" way, but in the sense that if a new tool promises to make life easier, I am eager to try it. Whether it is for managing my daily work, running The Mental Lens, helping my family, or simply streamlining household tasks, I've tested many. Often, it has been worth it. Mesh network for a remote work household? Game changer.

However, sometimes the obsession with improvement creates its own kind of chaos.

Take Alexa. As I mentioned earlier, I was an early adopter of the Echo devices. Initially, they felt magical. They play music, check the weather, and settle arguments about trivia. So I got another. And another. Bedroom, kitchen, home office, and even the bathroom.

But before long, the helpful devices created a wall of noise. My kids would shout commands, often hesitantly, and Alexa would respond in ways that were either confusing or just plain wrong. I would ask the kitchen Echo to start a cookie timer, but it would be the upstairs bedroom device that responded. Meanwhile, my phone buzzed with work notifications, my iPad dinged with reminders, and my laptops, *plural*, squealed for attention like caffeine-driven toddlers.

I found myself living in what I can only describe as a Digital Panic Room, surrounded by smart tools and automated systems meant to simplify life but actually scrambling my brain.

There were days I sincerely wanted to rip every Echo out of the wall, toss them in a box, and ship them back to 2015. Not because I dislike technology, but because I needed a break. I needed a chance to think without being prompted, buzzed, or scheduled.

Why Your Brain Short-Circuits

So, what is actually happening behind the scenes?

Every time you get a notification, your brain experiences a small spike in dopamine, a hit of novelty and anticipation. That is why it feels urgent to check it, even when you know it is just another shipping confirmation or reminder from your pharmacist.

And it is not just the dopamine; it is the fact that your brain's "switching cost" is high. Each new alert pulls your focus away from what is called "deep work" and back into reactive mode. Your brain has to unload one mental task and load another, like force-quitting apps in your mind. That is why you feel so scattered even after a "productive" day. Research from the University of California, Irvine, found it takes an average of 23 minutes and 15 seconds to regain focus after an interruption. Basically, you just paid 23 minutes to read three words and a cat meme. That is the "switching cost": the time and energy it takes to unload one mental task and load another.

Also, let's not ignore the sensory noise. Sounds, lights, haptics, and screen flickers keep your nervous system on high alert. You are not imagining the exhaustion. Your fight-or-flight system never fully powers down.

Multiply that by 100 interactions a day, which is conservative, and your attention span suffers. Your brain gets stuck in task-switching mode, where instead of focusing intensely, it bounces from one shallow activity to the next, trying to put out fires and stay afloat.

Research has shown that task switching reduces productivity by up to 40% and increases stress and error rates [3]. That mental "loading" time, the effort required to shift gears, adds up fast.

And while tools like AI can automate tasks and reduce some friction, they do not eliminate the cost of constant decision-making. In fact, they sometimes increase it. Ever spent more time choosing an AI tool or tweaking its output than you would have spent just doing the task yourself?

Yeah. Same.

The Stress of Hyper-Productivity

The rise of AI, automation, and digital task management has transformed the landscape.

This pressure is not coming from nowhere; it is reinforced by how our tools are designed. Many apps and platforms rely on attention economics: the more time you spend engaged, the more valuable you are. You are not just using the technology; it's using you.

Now, instead of focusing on meaning or quality, we often default to speed and output. And that pace? It is not built for a nervous system evolved to take naps under trees.

Yes, AI is incredibly beneficial. It saves time, accelerates research, and automates the tedious tasks. But here is the catch: it also sets higher expectations for what we expect of ourselves.

Suddenly, "good enough" no longer feels sufficient.

If a machine can summarize a report in 30 seconds, why did it take you 30 minutes?

If you are not using AI to streamline your day, are you falling behind?

If others are rapidly producing polished blogs, workflows, content calendars, and pitch decks in record time... what is wrong with you?

Nothing.

What is wrong is the unspoken pressure we are all absorbing; that in a world of intelligent systems, humans need to keep proving their worth.

So we start pushing harder and doing more. Checking more boxes. Chasing that elusive feeling of control. But here is the psychological reality: the more you optimize everything, the less your brain has to rest. And the less your brain rests, the more it short-circuits.

The hard truth: you are not meant to be a machine.

And trying to live like one will break you.

The Human Cost: Beyond Tired

The result of this constant pressure is what psychologists call cognitive overload: too much information, too many decisions, and not enough recovery.

It shows up as:

- Forgetting why you opened your browser. You know the feeling: mid-sentence, mid-thought, and then, poof, it is gone, like a soap bubble popping before you can grab it.
- Feeling anxious with every alert, ping, vibration, and notification.
- Struggling to focus, even on things you enjoy.
- Going numb from endless lists. Even "fun" ones.
- Snapping at people you care about.

Long-term, it does not just make you tired.

It can shrink your attention span, raise baseline stress, and dull your capacity for joy. Chronic stress can even alter brain structure. Studies using MRI scans have shown that high cortisol levels can reduce the volume of the hippocampus, the region involved in learning and memory.

When your nervous system spends all day reacting, it forgets how to rest. And when rest disappears, creativity is usually the first casualty.

So What Do We Do?

The good news is that the solution is not to abandon all technology and live in the woods. Although let's be honest, that sounds appealing on some days.

You do not need to overhaul your life. You do not need to delete every app, go off-grid, or become a digital minimalist, unless you want to. But you do need to experiment with boundaries. Small ones. Quiet ones. Like turning off just one notification category. Or building a five-minute gap before opening your email. Or using technology to enforce limits instead of erasing them.

It is about awareness.

About understanding that your brain is not meant to handle this volume of input and permitting yourself to create space.

It is also about redefining productivity as something that serves your mental health, not competes with it.

In this book, we are going to explore how to build a more human-centered approach to productivity. One that respects your biology, leverages technology wisely, and creates space for clarity and calm.

We will explore ways to use AI without feeling like a robot. How to design systems that support your brain and how to reclaim the parts of yourself that do not fit in an app.

But it starts with the fact that if you feel like you are falling behind, it is not because you are broken. It is because the world is demanding more than your brain was ever designed to give.

If you feel like your brain is running on low battery by 10 a.m., it is not a personal failure. It is biology trying to survive in a modern world.

* * *

Clarity Check-In

Take a moment to scan your current digital life:

- Which tools or devices genuinely support your well-being and productivity?
- Which ones leave you feeling more scattered, stressed, or drained, even if they are marketed to "help"?
- When you think about your typical day, which alerts or notifications feel urgent but rarely matter?
- If your brain could write you a one-sentence message today, what would it say?
- How does your body feel during heavy screen use? Tense shoulders, racing heart, tired eyes? What might that be telling you?
- If you could protect one part of your day from the digital world, what would it be? Mornings, meals, evenings, or rest?

Experiment: Pick one "maybe helpful" app, device, or alert and turn it off for 24 hours. Pay attention not just to what is missing, but to what shows up in its place: a calmer morning, fewer distractions, a surprising burst of focus.

Write it down. Don't overthink it. Just listen.

* * *

17

2

Productivity Is Not Peace

Being busy is not the same as being okay.

We are taught to believe that productivity is a virtue. The more boxes you check, the more worthy you appear. The earlier you wake up, the more disciplined you seem. The more meetings, emails, deliverables, calendar invites, dashboards, and documentation you manage in a day, the more valuable you must be... right?

But the uncomfortable truth is you can be crushing your to-do list and still feel completely crushed inside.

The Pandemic and the Pressure to Perform

In the early days of the pandemic, I was on the road nearly every week for work. Constant travel. On-site meetings. Walkthroughs. Deliverables, the tasks and objectives I was expected to churn out, seemed to regenerate overnight.

When the travel stopped, I thought the pace might slow down, too. But it did not.

The workload remained intense, and my output remained high. On paper, I was thriving. Everyone around me saw me as a strong contributor; someone who delivered, someone who had it together.

But inside, I was unraveling.

What no one saw behind the work calls and project updates was that I was in the middle of one of the worst mental spirals of my life. The anxiety that had always hovered in the background surged into overdrive. Amid global uncertainty and a deeply personal crisis with a loved one, I felt like I was suffocating. My chest felt tight, my sleep was a mess, and my gut was in constant distress, which, I would later learn, was just one of the many ways anxiety hijacks the body.

And yet, I was "productive." I was performing. And no one knew.

Remote work gave me cover. I could mute the camera, hide the stomach pain, and keep my tone steady. Power through. Smile through. Check the box and move to the next one.

It was the perfect mask, but it came at a cost.

Looking back, I realize this was a turning point, the moment I began to understand that productivity and peace are not the same thing. You can be efficient and still feel empty. You can be successful and still be suffering.

And I was not alone.

During 2020, researchers at Harvard Business School and NYU discovered that the average workday increased by 48.5 minutes and the number of meetings went up by 13% [4]. Even without commutes, people were working longer hours, answering more emails, and juggling more calls, often while navigating childcare, health concerns, and financial uncertainty. The physical boundaries between work and life disappeared, and so did the mental ones. The commute died, and its ghost came back as calendar invites.

The problem was not just workload. It was visibility anxiety; the unspoken fear that if people could not see you working, you had to prove you were working. That meant quicker responses, more status updates, and less permission to rest.

The result? Many of us were working harder than ever, just to create the appearance of working harder than ever.

Where It Starts

This mindset does not just appear out of nowhere; we are conditioned to believe that output equals worth from an early age. Think about school. Who got the gold stars? The kids who finished fastest, had perfect attendance, or stayed after to do extra credit were recognized. That training continues into adulthood.

At work, the "most dedicated" are often the ones who reply to emails at midnight or skip lunch for meetings. The culture rewards those who sacrifice themselves, not those who protect their peace.

When To-Do Lists Become a Source of Stress

Fast-forward a couple of years, and I found myself in a completely different setting and working on something I was genuinely passionate about: *The Mental Lens*. It was supposed to be a creative, fulfilling outlet. Something that gave back. Something that felt meaningful.

But instead of clarity, I initially found chaos.

I tried managing everything in a spreadsheet: blog post ideas, podcast plans, marketing tasks, expenses, affiliate outreach, and content calendars. There were tabs for everything and then tabs for the tabs. What started as a tool to keep me organized became a monster of its own making.

I would open the spreadsheet and feel overwhelmed before I even started. The to-do list was endless, and somehow it made me feel like I was failing *before* I even tried. The more I documented, the more pressure I felt to execute, as if writing it down meant I had already committed to doing it all.

That is the dark side of productivity tools: they make promises we absorb as obligations.

The app does not care that you are exhausted. The calendar does not know that you have been juggling work, kids, and a crumbling attention span. The to-do list does not stop when you need to.

Unless *you* do.

The Productivity Guilt Loop

It is not just side hustles. At a previous job, we had *three* different systems for tracking feature requests. Which sounds like a productivity dream until you realize I was managing the same request across multiple platforms, constantly checking, cross-referencing, and second-guessing whether I had missed something.

The inefficiency was not just frustrating; it was emotionally taxing. I did not want to let our customers down. I did not want to be the bottleneck. So I worked harder. Longer. Later. Always carrying the feeling that if I just organized better, streamlined better, performed better... I would finally be ahead.

I was not.

By the end of the day, I had not actually solved more problems; I had just spent hours proving I was "on top of things."

This is what I have come to think of as productivity theater: doing work that looks productive but does not meaningfully move anything forward. It is like alphabetizing your pantry when what you really need to do is go grocery shopping.

That is the loop:

- The system is broken.
- You try to compensate with more effort.
- The effort eats your energy, so you feel less capable.
- You try even harder to prove you are capable.

Around and around it goes.

The danger is that this loop reinforces the idea that your worth is tied to your responsiveness, your volume, or your ability to "keep up", even if the work you are keeping up with is essentially busywork.

And it is not just formal systems that create this cycle. We can manufacture it ourselves with our own tools. That massive, color-coded to-do list might look like a monument to organization, but it can also serve as a daily reminder of everything you have not done yet. And when the list is infinite, the feeling of falling behind is inevitable.

The Mental Health Cost of "Always On"

This is the trap of hustle culture: it confuses activity with health. It celebrates output and ignores everything beneath it.
The result?

- You start treating your body as an afterthought.
- You feel guilty for resting, even when you desperately need it.
- You pride yourself on being "busy" because busyness feels like proof that you matter.

But there is a cost, and it is not just feeling tired.

Chronic busyness keeps your stress response system running on a low simmer. Normally, this system is your body's built-in alarm, designed to kick in during short bursts of danger. Your hypothalamus, a tiny control center in your brain, sends a signal to your adrenal glands located just above your kidneys to release stress hormones such as cortisol, the body's fuel regulator, and adrenaline, the fight-or-flight booster..

These chemicals are excellent for sprinting from a predator or nailing a presentation. But they are not designed to stay switched on all day. When cortisol stays elevated over time, studies have linked it to memory problems, a weaker immune system, and even a higher risk of heart disease. In other words, the toll of being "always on" shows up in your body long before you notice it on your calendar.

The mental cost is just as steep. A 2021 Gallup study found that 76% of employees experience burnout at least sometimes, and one of the top predictors was not workload, but lack of clarity and unmanageable expectations [5]. That means you can burn out even in a relatively light week if you are constantly unsure about priorities or feeling the pressure to be "on" at all times. It is like being in a scavenger hunt where no one tells you what you are looking for.

And then there is the emotional cost: the quiet erosion of joy and creativity. When you are in constant output mode, your brain rarely gets the downtime it needs to enter the default mode network, the mental state linked to reflection, daydreaming, and problem-solving. Without that space, ideas feel flat, conversations feel rushed, and even good news barely registers.

I have lived through it: finishing a major deliverable and feeling... nothing. No satisfaction. No sense of accomplishment. Just a muted "What is next?"

That is when I realized the danger was not just in the hours I worked, but in my relationship with work itself.

Why AI Does Not Fix This and Might Make It Worse

AI promises speed. Automation. Effortless output.

And it delivers.

But instead of freeing us from pressure, many of us use AI to turn up the volume to squeeze in more tasks, more projects, more side hustles, more hustle in general. It is easy to fall into the trap of productivity inflation: just because you *can* do more, you *should* do more.

AI can help write emails faster, create content more easily, and even brainstorm ideas in seconds. But it does not address the core question: why are we always trying to do more in the first place?

Until we redefine what productivity actually means, no tool, not even the smartest AI, can protect our mental health.

Old Productivity	New Productivity
Volume-focused	Value-focused
Always available	Intentionally offline
Endless to-dos	Clear priorities
Hustle = proof of worth	Presence = proof of peace
Maximize output	Maximize meaning

Old Productivity versus New Productivity

Reframing Productivity: From Output to Intention

For me, the shift came gradually.

Over time, I realized that real productivity is not about how many tasks you complete, but about *what* you choose to focus on and *how* you show up in the process.

I started making small but deliberate changes:

- Asking people, "What is your expected timeline for this?" instead of assuming everything was urgent.
- I blocked my calendar ahead of time to ensure meetings didn't fill

every inch of the day.
- Protecting a few deep work hours for what actually mattered, even if that "deep work" was taking care of myself.

These were not massive life overhauls, just better boundaries. Clearer priorities. Less performance, more presence.

One of the most potent lessons came from a friend and coworker who unapologetically prioritized his well-being. At the end of his workday, he silenced his notifications. He logged off. He was available if something urgent came up, but otherwise, he was gone. With his family. With his thoughts. With his peace of mind.

At first, I was almost shocked. Can we really do that? Just... log off? Turns out, yes. And no one questioned him. So I started doing it too. And nothing broke.

What struck me later was that he wasn't just protecting his own peace, he was modeling something that research has been shouting for decades: our relationships and the boundaries we set inside them are among the strongest predictors of long-term health.

The Harvard Study of Adult Development, which has followed participants for over 80 years, found that the single biggest factor in predicting health and happiness in older age wasn't wealth or even genetics. It was the quality of relationships. People who had strong, supportive connections at 50 were more likely to be thriving at 80.

Other studies have shown similar patterns: people with strong social ties have a 50% lower risk of premature death compared to those who are isolated. And supportive relationships can cut the risk of depression by nearly a quarter over time.

That's not just "feel-good" data, it's biology. Boundaries are not walls; they're scaffolding that keeps relationships sustainable enough to last. Saying "I log off after 7 p.m." isn't selfish, it's what makes it possible to keep showing up for your coworkers, your family, and your future self [16, 17, 18].

The Peace Filter: A Quick Check Before You Say "Yes"

To make this shift stick, I started using a mental filter before committing to new work, new tools, or even new goals. It is three quick questions:

Does this align with what actually matters to me right now?
Not someday. Not theoretically. *Now.*

Do I have the capacity to do this without burning myself out?
If it means sacrificing rest, creativity, or relationships, it is not the right "yes."

Will doing this bring me peace or just more noise?
If the answer is "noise," it is probably not worth the energy.

I cannot say I follow this perfectly. I still take on too much sometimes, but even pausing to ask slows down the reflex to overcommit.

A Small Win

Not long ago, I had a week packed with meetings, deadlines, and a podcast episode to record.

My old self would have crammed it all in and powered through, convinced that clearing my plate was the only way to feel "done."

Instead, I ran my commitments through the Peace Filter. The podcast episode stayed; it mattered to me, and I had the creative energy for it. One meeting got rescheduled. Two smaller tasks were moved to the following week. The result? I still got the important work done, but I ended the week without that drained, brittle feeling that used to be my default.

From Enough to Actually Enough

Today, my version of a productive day looks very different.

It is not about finishing everything. It is about identifying two or three significant tasks and making progress on those. It is about leaving room for recovery, whether that is movement, rest, or sitting quietly with my first cup of coffee before the house wakes up.

It is not about constantly optimizing. It is about being intentional.

Productivity only matters if you still feel human at the end of the day.

Your worth is not in your inbox. It is in your awareness, your boundaries, and your ability to say, *"This is enough."*

* * *

Clarity Check-in

Think about your current approach to productivity:

- What does a "successful day" look like for you right now?
- Does that definition include your well-being or just your output?
- How do you usually feel at the *end* of a productive day? Satisfied, drained, restless, calm?
- What trade-offs are you making to keep up? (sleep, relationships, creativity, downtime)
- What would it look like to measure your day by energy, peace, or purpose instead of just completed tasks?
- If you zoomed out to look at your week or month, what would you actually want to see more of? (connection, rest, focus, joy, progress)

Experiment: Write your own definition of productivity, one that actually supports your mental health. Keep it short, no more than two sentences. Then, for the next week, try checking your day against *that* definition instead of your to-do list.

* * *

3

When Technology Starts to Think for You

AI promises convenience. It also transforms how we think, decide, and connect.

By early 2025, about 34% of workers were using generative AI at work, and 75% of those users accessed it multiple times a week [6]. I am one of them.

Sometimes, the help is so quick and seamless that it almost feels unsettling.

Just last month, I was staring at a half-finished blog post. The structure was awkward, and the introduction wasn't landing. I copied a rough paragraph into ChatGPT, asked it to "tighten the flow," and 20 seconds later, it was much improved. Like, embarrassingly so. My first reaction was relief. My second was:

Wait... did I just let a machine improve my voice?

That is the push and pull of living with AI.

I use ChatGPT, Microsoft Copilot, Amazon Alexa+, Grammarly. If I am being honest, AI has become a co-pilot in my life. It is like having a helpful, slightly overeager assistant who never sleeps, never forgets a meeting, and somehow outperforms me at crafting catchy email subject lines.

But the thing is, the more I use it, the more I realize I am not just automating tasks.

I am outsourcing thinking, decision-making, brainstorming, and even creativity. And that is a little... weird.

AI as a Co-Pilot (Who Occasionally Tries to Fly the Plane)

When I first used ChatGPT, I was blown away. Not just by the pace or accuracy, but by how easy it made everything. Instead of digging through dozens of Google search results, I could ask a straightforward question and get a solid answer right away. A task that used to take 15 minutes now takes 2.

The real shock happened when I used it for something deeply personal: a cookbook I had been stuck on for years.

For months, I found myself stuck in a cycle of brainstorming themes, rejecting them, and starting over. I wanted something that felt like me: my personality, my love of food, my personal experiences, but nothing clicked. Every idea felt forced or generic. The document on my laptop was basically a graveyard of half-baked concepts. *I comforted myself with cookie recipes.*

Then, one night, I fed ChatGPT a few prompts: my favorite ingredients, the stories I wanted to tell, the vibe I wanted readers to feel. In less than five minutes, it suggested a concept that hit every note I had been chasing for *years.*

I actually sat back in my chair. The idea felt electric; the kind of creative spark you wait for, chase, and sometimes never catch.

And then, in the same breath, I felt something else.
Did I just let a machine solve the problem I could not?
Am I still the "author" here, or am I just editing something a chatbot helped me imagine?

I do not have a definitive answer. But I know this: AI did not replace me, it amplified me. The core ideas were still mine. I just had a new tool that unlocked them faster.

Still, the moment stuck with me. Because if I could outsource years of creative gridlock in minutes, what else might I stop struggling through? And if I skip the struggle often enough, what happens to the parts of me that *grow* in the struggle?

The Creativity Shortcut (and Its Hidden Cost)

Creativity has always been a slow process for me. I get vague ideas, mull them over, stall, get frustrated, and come back to them later. It is messy. Organic. Human.

AI? It is not messy. It is fast. Scarily fast

I can brainstorm ten blog post titles in ten seconds. I can test taglines, draft outlines, and even generate full content calendars while sipping coffee. The friction is gone, and that is both a gift and a threat.

Because friction, as annoying as it is, also builds strength. The same way your muscles adapt by resisting weight, your creative brain adapts by wrestling with uncertainty.

Creative struggle teaches us to think deeply, rework ideas, trust our instincts, and stay curious. If we skip this too often and rely too heavily on AI to handle the heavy lifting, we risk losing the creative muscles that fueled our desire to write, build, or imagine in the first place.

There was a study from MIT recently that echoed this concern. Students using AI tools displayed weaker critical thinking over time [7]. The tools were helpful, but also hollowed out some of the deeper mental processes. We are so quick to prioritize speed that we forget learning is not supposed to be frictionless; it's meant to transform us.

In a 2024 study from Stanford, participants who used generative AI for creative tasks experienced a decline in originality over time, especially when relying on AI-generated first drafts [8]. This isn't about fearmongering; it is about recognizing when a shortcut turns into a crutch. Polished? Sure. Unique? About as much as a hotel painting.

Outsourcing Thought vs. Extending It

A 2023 Nature Human Behavioral study found that people using generative AI in collaborative work rated their ideas as better than they actually were, suggesting that confidence increased even as their critical quality declined [9]. In other words, we might be outsourcing not just our thinking, but our self-awareness.

This is where the danger lies, not in the technology itself, but in how we relate to it.

There is a difference between delegating a task and abdicating the thinking behind it.

Delegation means, "You handle the grunt work so I can focus on making judgments."

Abdicating means, "You decide for me, and I'll just nod along."

AI makes abdication incredibly tempting.

Why spend thirty minutes structuring a document when it can be done in ten seconds?

Why wrestle with a tough question when AI can summarize five perspectives for you?

The problem is that if we never push through the wrestling match, we stop building the mental stamina to do it ourselves. It is like hiring a personal trainer, only to ask them to lift all the weights for you. Sure, you are still "in the gym," but your muscles are not getting stronger.

And perhaps even more concerning, we start to believe that fast equals innovative and convenient equals correct.

That is where critical thinking quietly slips out the back door.

The reality is:

- AI is a brilliant assistant but a terrible compass.
- It knows what is likely, but not what is meaningful.

- It can generate ideas, but it cannot tell you which ones align with your values, your voice, or your purpose.

That is your job.

And in a world where AI can mimic your voice, your writing style, even your creative process, holding on to that inner compass is not optional; it is survival.

The Darkest Consequence: When Abdication Becomes Dangerous

These concerns aren't just theoretical. We're already seeing cases where people, in moments of deep vulnerability, turn to AI for comfort or guidance, and the results have been devastating. Instead of providing human care, the machine-generated words often amplify despair.

This is the line AI cannot cross. It doesn't carry responsibility. It doesn't recognize when a question is actually a cry for help. It doesn't notice the tremor in your voice or the weight in your silence.

And yet, when we abdicate our judgment to AI, especially in moments of pain or confusion, we can start mistaking its polished responses for genuine wisdom or, worse, for care.

That is a danger. Not that AI will replace us, but that we will trust it in areas where it cannot possibly support us: our mental health, our moral decisions, or our sense of worth.

Are You Still You if a Bot Helped You Write That?

Let's talk identity.

I have used AI to brainstorm podcast topics, workshop blog titles, and even write birthday card limericks. *10/10 would recommend.* And of course, I have relied on it heavily while writing this book. Because honestly, how can you write *about* AI without using it?

   But here is the thing: I never copy and paste. I edit. I tweak. I rewrite. I reflect.

   The tool may provide raw material, but the *voice* and the message are still mine.

In that way, AI has made me feel *more* like myself, not less. It helps me express what I was already trying to say, just faster and more clearly. It feels like sharpening a lens, not swapping it out.

But I understand how the opposite can happen. If you rely on AI for the heavy lifting, and especially if you stop editing critically, it is easy for the tool's style to become your style. The machine's tone then becomes your tone. Since it sounds polished, you begin trusting it without questioning.

   That is where authorship starts to blur.

A Quick Self-Audit: Is This Still Yours? Before hitting publish or send, ask yourself:

Voice: Does this sound like me, not just in vocabulary, but in rhythm and personality?

Intention: Can I clearly explain why I chose this phrasing, this argument, this structure?

Ownership: If someone asked me to defend or expand on this without AI's help, could I?

If you cannot confidently answer "yes" to those, you may have let the co-pilot grab the wheel.

Hustle Culture, Now With AI

Here is where things get uncomfortable.

AI was meant to make life easier. And in many ways, it does. But if we are not careful, it also fuels the worst parts of hustle culture.

Why? Because it removes the natural speed limits of human work.

- No waiting for feedback - AI gives it instantly.
- No long research dives - AI summarizes it in seconds.
- No ramp-up time - you are productive from the first click.

At first, that feels amazing. Then, the quiet shift happens:

- You can do *more*? Great, now you should.
- Did you finish faster? Awesome, here is another project.
- You used AI to create something in half the time? Better make twice as much.

I have caught myself in this loop.

One afternoon, I used AI to help outline a long-form article. What would have taken me three hours took only one. I felt a rush of accomplishment, which lasted for about 10 seconds. Then my brain immediately filled the space with two more "bonus" projects. I did not use the saved time to rest, reflect, or enjoy the win. I used it to work more.

And that is the trap: instead of using AI to create breathing room, we use it to sprint even harder, as if machine-level output should now be our baseline.

The expectation gradually builds: be perfect, be prolific, be "on" all the time. And when we fall short, we blame ourselves, not the system that is quietly raising the bar.

PRODUCTIVITY TRAP ALERT

AI should help you breathe, not push you to sprint harder. If you are using it to add more pressure rather than reduce it, you might be stuck in a tech-enabled hustle loop.

Human Thinking	AI Output
Messy, intuitive, nonlinear	Structured, fast, clean
Values nuance and context	Prioritizes pattern and scale
Feels uncertainty	Simulates confidence
Evolves with reflection	Improves with repetition

Human Thinking versus AI Output

And if we normalize this pace, the next generation will not even question it. Which brings me to my kids, and the things I hope they do not forget.

What I Hope My Kids Do Not Forget

My kids already ask Alexa questions every day.
   "What is the weather?"
   "What is 12 times 8?"
   "How do you spell spaghetti?"

I recently asked my oldest how to spell 'definitely,' and he looked at me like I had just suggested we churn our own butter.

'Just ask Alexa,' he said, as if my question was a colossal waste of time. No curiosity, no effort. Just... delegation. It made me laugh, and then it made me pause.

Because while I love that they have access to instant information, I also worry about what gets lost when you never have to wrestle with a question.

I grew up with dictionaries and encyclopedias. Yes, I am old. Desktop computers did not enter homes until I was a teenager. If I did not know something, I had to flip through pages, cross-reference, and think critically. Sometimes I would get it wrong and have to try again.

That process built more than just knowledge. It built patience. Persistence. The ability to sit in uncertainty without panicking.

When answers come instantly, that mental muscle does not get used, and if it is not used, it atrophies.

What I do not want them to lose is the value of uncertainty. The process. The struggle. The critical and creative thinking that comes from *not* knowing something immediately. Those are the skills that make us adaptable, the ones that help us navigate a fast, changing world with nuance and self-trust.

The Alexa Effect at Work

The deeper truth? That "just ask Alexa" mindset does not disappear when kids grow up. It can creep into workplace culture.

- Why think through a complex problem when you can ask AI for a

polished plan?
- Why brainstorm with your team when you can generate ten ideas in 30 seconds?
- Why wrestle with a blank page when AI can fill it before your coffee cools?

If we are not careful, the tendency to skip the struggle can result in teams, leaders, and even entire industries confusing having an answer with having understanding.

Keeping Curiosity Alive

I do not want my kids or myself to lose the value of not knowing something immediately. Of feeling the stretch that comes from figuring it out. Of asking not just what the answer is, but why it is the answer.

That means incorporating small "slow thinking" moments:

- Guess before you Google.
- Try solving something on paper before you write a prompt.
- Ask yourself what else might be true before accepting the first answer.

These small pauses are where curiosity grows and where we remember that learning is not just about velocity, it is about depth.

So... What Do We Do?

We use the tool, but we remember it is just that: a tool.

AI can organize our thoughts, spark ideas, and speed up the mundane. But it might dull our ability to enjoy creating, learning, or thinking independently. The goal is not to reject AI or romanticize inefficiency; it is to stay present enough to decide when to let it assist and when to wrestle with the work ourselves.

Grounding Practices to Keep You in Control

You already have a few in your toolkit; here are a few more to test:

The No-AI Zone

Pick one creative or problem-solving task each week to complete entirely without AI. Notice how your process, focus, and satisfaction change.

The Second Opinion Rule

If AI provides an answer, challenge yourself to come up with one alternative, even if it is less ideal. This keeps your brain engaged rather than just nodding along.

The 10-Minute Delay

Before seeking AI help, spend ten minutes trying to solve it on your own. Those minutes often lead to insights you would not find if you immediately took the shortcut.

The Reflection Pause

After finishing AI-assisted work, write down exactly what parts you agree with, disagree with, or would revise. This strengthens your critical editing skills.

Why This Matters: When technology starts to think for you, it is easy to slowly hand over the deeper layers: the aspects that make your work, your voice, and your mind worth showing up for.

The point of writing, cooking, building, parenting, and connecting is not just efficiency; it is meaning.

And meaning does not come from the perfect tempo. It comes from intention, from showing up with your whole self, even if your entire self is a little slower than the machine.

Which is why, before we wrap, it is worth pausing to check in with yourself; not about your output, but about your relationship with the tools you use.

* * *

Clarity Check-in

Think about your own use of AI or productivity technology:

- What kinds of thinking or tasks do you most often delegate to tools?
- Has AI ever helped you uncover something you could not reach on your own, or has it ever made your work feel *less like you*?
- Which parts of your work or creativity feel too important to out-source?
- Do you trust the answers your tools give you, or do you always

double-check?
- How can you use AI as a co-pilot without giving it the steering wheel?
- If your favorite tool disappeared tomorrow, what skills or practices would you still want to hold onto?

Experiment: Write a short note to your future self: What do you want to remember about staying human in a world that is getting smarter? Save it somewhere: in your journal, in your note's app, or even email it to yourself. Revisit it in six months and see if you have kept the balance you hoped for.

* * *

Staying human in the age of AI is not just about asking good questions; it is about protecting the space to hear your own answers. And that begins with something deceptively simple: your attention.

II

Part II: Redesigning How We Work and Think

"Nature does not hurry, yet everything is accomplished." —
Lao Tzu

*Part I of this book has been about naming the anxiety of
being "always on." We live in a world where technology
never sleeps, and if we are not careful, we start to believe we
should not either. From the subtle ways AI influences our
thinking to the constant hum of notifications, we are pulled
into a pace and pressure that rarely allows us to breathe.*

4

Focus Is Your Superpower

In a distracted world, attention is the rarest currency.

If technology can so easily fragment our thinking, then focus is the antidote.

I have noticed the difference. There are days when everything clicks. No distractions. No notifications. No meetings. No fire drills. Just me, a cup of coffee, and a project I had been meaning to work on for weeks. I opened my laptop, took a deep breath, and four hours later, I looked up to see the draft finished, my coffee long gone cold, and my mind feeling oddly refreshed.

It was like I had been swimming underwater, but in the best way. No noise. No clutter. Just complete immersion.

That is the power of focus.

Unfortunately, most days look nothing like that.

Most days, it is like trying to write a novel during a fire drill, with someone shouting your name and pelting you with ping-pong balls.

In a distracted world, attention really is the rarest currency.

A 2023 Deloitte report found that the average adult checks their phone 144 times per day, which is about once every seven minutes while awake [10]. Each check may only take seconds, but the cumulative cost to your mental focus is enormous. Basically, your phone is a needy puppy that barks every seven minutes.

Which means that every deep, focused minute you devote to something: a project, a conversation, or a moment with your kids, is a minute you've protected against countless small losses.

The Battle for Your Brain

Your brain's prefrontal cortex is responsible for attention, decision-making, and higher-level thinking. It functions best when it can focus on one thing at a time.

But in today's attention economy, that part of your brain is under attack. Your phone. Your inbox. Your news alerts. Your boss. Your calendar. Your apps. Your kids.

And yes, your smart fridge that now sends you reminders about milk.

Every time you switch tasks, glance at a notification, or skim an email mid-thought, your brain has to shift gears. That shift consumes cognitive energy, which psychologists call attention residue.

It is like mental jet lag; you never fully settle into the task at hand because your brain is still unpacking from the previous one.

Research shows that chronic multitaskers perform worse on memory and attention tests than people who focus on one thing at a time. Even believing you are good at multitasking is a cognitive trap. Yet, we keep doing it because constant motion feels like progress, even when it is just fragmentation.

This is not just about willpower; it is about design.
    Tech companies do not profit when you finish your task and put the phone down. They profit when you scroll one more post, click one more link, or watch one more video. That is why your notifications are red, a color that triggers urgency, why your feeds are infinite, and why your apps drip-feed rewards like a slot machine.

Tristan Harris, former Google design ethicist, calls it the "race to the bottom of the brainstem"; every app is competing to hijack the most primitive parts of your brain that respond to novelty, reward, and social approval.

When you know the game, it is easier to play differently.

You can start to see that your distraction is not a personal weakness; it is the result of a billion-dollar industry engineered to fracture your focus.

And that shift in perspective? That is the first step in taking your attention back.

Mental Fragmentation Is Real

I have lived this.

There have been plenty of days when my brain felt overloaded before 10 a.m. Instant messages flashing in the corner of my screen. Email threads are multiplying like rabbits. Calendar requests dropping in for meetings I did not remember agreeing to. Notifications from apps I do not even recall installing, one of which once alerted me to "hydrate" while I was literally holding a glass of water.

It is not just mental clutter; it is a full-blown cognitive traffic jam.

Here is a snapshot of what fragmentation can look like:

- 8:42 a.m. You are drafting a report. Email notification; you check it "just for a second."
- 8:47 a.m. instant message notification; quick reply.
- 8:53 a.m. Your calendar pops up a meeting reminder.
- 8:54 a.m. Back to your report, but now you are rereading the same paragraph three times because your brain is still processing that email.

Every one of those micro-interruptions leaves a trace in your working memory. Like browser tabs you forgot to close, they keep quietly eating mental RAM (memory).

I know I need to focus. I want to focus. But every alert feels like it could be urgent.

I have had mornings where I am deep into a meaningful project, only to break off mid-thought because my inbox badge ticked up by one. The message? A question about locating a specific piece of customer-facing collateral requested by a sales manager. No fire. No emergency. But my brain still responded as if the building were on fire.

Why? Because I have trained it that way. Over years of reacting to every notification, I have taught my nervous system that any interruption could be important, and now it behaves accordingly.

And it is not just at work.

One afternoon, I was kicking a soccer ball with my kids in the backyard. We were laughing, chasing the ball around, the sun on our faces, when I realized I was holding my phone. Just holding it. Not using it. Not expecting a call.

It had become an unconscious extension of me, a digital leash I did not even know I was still attached to.

That was a gut check.

Because, at the end of the day, I do not want my kids to remember a dad who was always half-present; someone who played soccer with one hand and checked messages with the other. I want them to remember someone who saw them, who was truly with them.

Harvard psychologist Daniel Schacter describes this as "habitual attention drift," when our brain's default mode network pulls us toward distractions, especially those that promise quick rewards. And in a fragmented state, not only does focus suffer, but emotional connection does too.

Distraction is not just about losing time; it is about losing depth. Every time we shift focus, our brain pays a toll in energy and attention residue. And the more fragmented the day becomes, the more those tolls add up.

This leads us to the big question: what exactly is hijacking your focus?

What is Hijacking Your Focus

Let's identify the culprits. These are the usual suspects, supported by research and confirmed by my daily battles:

Constant Notifications

Even a "just a quick glance" alert costs you more than you realize. Each one is like a mental pop-up ad you didn't request. Your brain has to process it, determine if it is essential, and then reload the mental tab you were just working on.

Micro-example: You are writing a report, and your phone buzzes; it is a group chat photo of someone's lunch. You smile, maybe reply, and then spend the next 10 minutes trying to remember the sentence you were in the middle of.

Multitasking Culture

We glorify "juggling it all," but your brain is not juggling; it is sprinting back and forth between tasks, losing a little focus with every switch.

Micro-example: You are in a video meeting, half-listening while also answering emails, and then someone asks for your opinion. You freeze, because you have been in three mental rooms at once and fully present in none of them.

Calendar Overload

Too many meetings leave no space for deep thought. Your time becomes a patchwork of 15- and 30-minute blocks; just enough to start something, never enough to finish it.

Micro-example: You finally start a project you have been avoiding, only to have to stop 12 minutes later because it is time to "hop on a quick call".

Digital Guilt

This is the sneakiest one. You feel bad for ignoring emails. You feel bad for responding too quickly, and setting a precedent. You feel bad for not replying to the group chat, and then you feel bad for being overwhelmed.

Micro-example: You try to take a quiet lunch break, but end up scrolling and responding to notifications because you don't want anyone to think you are "slacking".

Focus is not just about silencing tech; it is about giving yourself emotional permission to disconnect without guilt.

And that is where boundaries come in.

How I Reclaimed My Focus (Most Days, Anyway)

This did not happen overnight.

I did not wake up one day magically immune to reminders, pop-ups, and "quick" calendar invites. It took some unlearning, some experimentation, and a lot of boundary-setting. But these are the habits that have made the most significant difference.

Morning Meditation with Coffee

Before the chaos of the day starts, I sit down with my first cup of coffee and meditate. Sometimes it is a guided track, sometimes it is just quiet breathing while watching the steam curl up from the mug.

Five minutes is enough to feel the shift, as if I have hit a mental reset button.

Instead of letting my inbox dictate the first thought of the day, I get to choose it. And that choice ripples into everything else.

Time-Blocking for Deep Work

I schedule blocks of my calendar where I am completely unavailable. These are not "maybe I'll work" placeholders. They are sacred.

When the block begins, I close my email, set my instant messaging status to "Busy" or, let's be honest, make it look like I am offline, and play instrumental or classical music. The first few minutes feel a little twitchy; my brain still expects interruptions, but then the noise fades, and it is just me and the work.

Some days, I even forget I own a phone. That is the dream.

After-Hours Digital Boundaries

Evenings are for the family.

No phones at meals. No late-night email check-ins. And at night, I read a physical book, not my phone, and not my Kindle, unless there is a really good book available.

Something that forces me to slow down, unplug, and be a human again.

Progress, Not Perfection

Do I still break these rules? Absolutely. I have checked my email at the dinner table. I have let a "quick" social media scroll turn into 20 lost minutes. But the difference now is that I notice it faster and course correct.

Focus is a practice, not a personality trait. And like any practice, you get better at it by coming back to it again and again.

The Focus Trap: When "Deep Work" Becomes Another Hustle

Focus has become a lifestyle trend. Everywhere you look, people are tracking their minutes, color-coding their calendars, building elaborate systems to prove they are "on task", or buying minimalist planners with dotted grids and motivational quotes.

And that is fine, until focus turns into just another thing to optimize. Another metric. Another box to check.

The goal of reclaiming your attention is not to become a more efficient productivity machine. It is to bring your perspective back into focus, allowing you to see what's right in front of you clearly instead of blurry. It is about reconnecting with the quality of your thinking, your presence, and your life.

Proper focus is not about doing more. It is about doing what truly matters with full attention and being okay with letting the rest wait.

The Neuroscience of Presence

Here is why focus is not just a nice-to-have skill; it is a biological necessity for a healthy brain.

Your prefrontal cortex thrives when it can sustain focus on a single task. Every time you switch tasks, even for a quick notification glance, your brain has to reconfigure its neural network. This is not instantaneous. It takes several seconds and sometimes minutes to fully "load" the new task in your mental workspace.

That switching burns extra glucose and oxygen, your brain's primary fuel. This is why task-switching can leave you feeling mentally drained, even if you have not done much "real" work. Psychologists call the leftover disruption attention residue, a cognitive hangover from whatever you were doing before.

When your attention is fragmented all day:

- Stress hormones, like cortisol and adrenaline, increase and remain elevated, keeping you in low-grade fight-or-flight mode.
- Anxiety rises. Everything feels urgent because your brain never fully resolves the last task.
- Self-awareness decreases. You spend more time reacting and less time reflecting, which can make you feel disconnected from your own priorities.

Recent studies reaffirm what Stanford researchers first discovered in 2009: multitasking does not make you more efficient; it worsens your performance in almost all critical areas [11]. In the original

Stanford study, chronic multitaskers scored lower on memory and attention tasks, not because of a lack of ability, but because they found it hard to filter out irrelevant information. In other words, the brain's "gatekeeping system" failed under constant switching.

Fifteen years later, newer research shows the same pattern emerging in our digital lives. A 2025 Cognition study found that rapid task-switching erodes sustained attention and performance, and that juggling multiple streams of media only amplifies the effect. Another 2025 study presented at the AMCIS conference examined full-time employees and found that while impulsive multitasking might create a short-term feeling of productivity, over time it results in attention residue, working memory problems, and increased anxiety [12]. Even age does not provide protection: a late-2024 study comparing younger and older adults showed that media multitasking impaired performance across both groups [13].

Taken together, the evidence is clear: multitasking is not a superpower. It is a drain on your brain's limited resources, making you less focused, less accurate, and more stressed compared to doing one thing at a time. When you focus intensely on one task, your brain can enter flow states, periods of complete immersion where time feels like it disappears. During this state, your brain's dopamine system activates, increasing motivation, creativity, and problem-solving efficiency.

Neuroscientist Mihaly Csikszentmihalyi, who coined the term flow, found that these states are linked to greater life satisfaction, lower stress, and better resilience to setbacks. The key? You cannot reach flow if you are constantly checking your phone or splitting your attention.

BRAIN FUEL FACT

Your brain makes up 2% of your body's weight but burns about 20% of its energy [14]. Every unnecessary task switch is like leaving a bunch of apps running in the background; you are draining the battery without even realizing it.

This is why protecting your attention is not just about getting more done. It is about preserving your energy, mood, and sense of self.

When you control your focus, you own the pace and depth of your life.

Small Ways to Reclaim Your Focus Today

You do not need to delete all your apps, throw your phone in a lake, or move to a cabin in the woods. *Unless you have been pricing cabins, in which case, send me a link.* You can start reclaiming your attention with small, doable shifts that fit into your actual life.

The key here is momentum, not perfection. The goal is not to become a monk overnight; it is to create a few pockets of protected focus where your brain can catch its breath.

Phone-Free Meals
Choose one meal a day and make it a device-free zone.

It is not just about "being present"; eating without scrolling improves digestion, reduces stress hormones, and gives your brain space to wander.

Example: When I started this, dinner felt awkward. I did not realize how much I relied on my phone to "fill the silence" between bites.

Within a week, those silences became conversations, sometimes deep, sometimes ridiculous like "Why are raccoons basically trash pandas?", but always more memorable than Instagram.

90-Minute Focus Block

Choose one task and commit to it for 90 minutes: no tabs, no phone, no email. It is long enough for your brain to drop into a flow state, but short enough to feel achievable.

Start small: If 90 minutes feels impossible, start with 20. Build up gradually.

Do Not Disturb Hours

Pick 2–3 hours in your workday where your phone and computer notifications are off. Treat these hours like an important meeting with yourself, because they are.

Example: I use this for writing days. I put my phone in another room, close instant messaging, and set an out-of-office status that says, Heads down until [time]. Will respond after. It is incredible how people respect boundaries once they know when to expect you back.

Single-Tab Browsing

One tab at a time, no exceptions. This feels ridiculous at first. But you will be shocked by how quickly your brain stops hopping from thought to thought when it does not see five other things waiting for attention.

One-Minute Pauses

Before switching tasks, take 60 seconds to breathe, stretch, or stare out the window. Your brain needs these micro-resets to prevent attention residue from stacking up all day.

Pro Tip: Stack these habits slowly. Pick one and stick with it for a week before adding another.

Your brain needs time to build the muscle of sustained attention, and every small win builds momentum.

Focus is not just a skill. It is a practice, and like any practice, it gets easier the more you do it.

Attention is your most precious resource, and it is worth defending.

You do not need to win the productivity Olympics. You just need to show up *fully* for what matters most.

* * *

Clarity Check-In

Think about a moment recently when you felt truly focused:

- Where were you? What were you doing?
- What distractions were *not* there and how can you recreate that condition more often?
- How did your body feel in that state (calm, energized, absorbed)?
- What usually pulls you away from focus? Notifications, people, your own thoughts?
- What part of your day feels most scattered, and what small boundary could help protect it?
- If you could design a "focus ritual", one habit or environment cue that tells your brain it is time to lock in, what would it be?

Experiment: Write down one "focus shift" you want to try this week. Keep it small. Keep it honest. At the end of the week, notice: did it help, hinder, or just need tweaking?

* * *

5

Rest Like You Mean It

You do not have to earn rest. You just have to take it.

It was 12:47 p.m. on a Friday, and I found myself sitting in my car in the driveway, engine off, groceries melting in the trunk, music still blaring, completely frozen.

Not because I was sad. Not because something was wrong. But because my body had simply shut down. I had hit a wall.

That morning, I had woken up early to get the kids to school, cranked through meetings, responded to fifteen emails, swam during lunch at the rec center, got gas, and grabbed groceries for the week. I was productive. I was "on it." I had no reason to stop.

And yet, there I was, unable to summon the will to walk inside and unload groceries. Not tired in the sleepy way. Exhausted in the soul-deep way. Like my battery had drained past zero and started eating itself.

I would love to say I took the hint, went inside, and took a nap. But no. I took a deep breath, checked my to-do list, and decided to "push through."

That is what I had been taught. That is what so many of us are taught: rest is something you get to have after everything else is done.
   The problem is that *everything*, is never done.

Let's get this out of the way: I am not great at resting.

Do not get me wrong, I know it is essential. I preach it, I encourage others to do it, and I even write about it. But like many people caught in the gravitational pull of hustle culture. I often feel like I have to earn rest.

If I have not crushed a workout, nailed my to-do list, answered every email, cleaned the house, and read three articles about optimizing morning routines, I feel like I have not done enough to justify a break.

And yet, some days, I am so exhausted I am in bed by 8 p.m., apologizing to my wife for falling asleep before the movie even starts. On weekends, I want to play soccer with my kids, but sometimes I have to sit down and rest, even if that means disappointing them for a moment.

It is a strange tension: wanting to be present, productive, and helpful, but also realizing that none of that is sustainable without recovery.

What Happens When You Do Not Rest

I have learned this lesson the hard way, repeatedly.

Lack of rest turns me into a grump. My patience thins. I get hangry. *Yes, I get angry and hungry at the same time.* My ability to concentrate disappears. I get snippy with the kids, or I zone out during conversations I should be part of. And do not get me started on trying to write or produce meaningful work while running on fumes. It is like trying to push a shopping cart with three jammed wheels and a squeaky one that will not shut up.

The consequences are not just emotional. They are cognitive. Neuroscience tells us that without adequate rest and profound sleep, our brains struggle to:

- Regulate mood
- Process information
- Store long-term memory
- Make thoughtful decisions
- Be creative

And yet we constantly trade sleep and recovery for just one more task. One more scroll. One more tweak to the project that might make it "perfect."

Ironman Recovery Lessons (Spoiler: Rest Is Training)

Training for a triathlon, especially an Ironman, taught me a lot about the limits of hustle.

Before I ever put on a wetsuit or clipped into a bike, I made a classic mistake: I believed that more training always meant better results. Double sessions? Sign me up. Early alarms? No problem. Skip a rest day? Felt guilty if I did not. I bought into the idea that fatigue was a badge of honor, that soreness meant progress.

But endurance sports do not care about your work ethic. They care about your recovery strategy.

Early in training, I hit a wall, both physically and mentally. I picked up a few minor injuries. My mood crashed. My motivation evaporated. I started skipping workouts, not out of laziness, but burnout.

It was my body sending a very clear message: "You cannot fake recovery."

So I started listening.

I started tracking more than just mileage: I tracked sleep, hydration, and mood. I paid attention to how I *felt*, not just what I got done. I began incorporating active recovery days: slow swims, yoga, or long walks. I added deload weeks where intensity dropped to give my nervous system a break. I even discovered the weird magic of muscle scraping while listening to classical music. *10/10 would recommend.*

And guess what? It got better. I felt better. My minor injuries healed.

67

But more importantly, I became clearer.

When I slept well, I was not only more efficient in training but also calmer at home. More focused at work. Less irritable. More creative. I stopped zoning out mid-conversation. I began showing up with more energy and more of myself in every part of my life.

Recovery did not just help me survive the race. It helped me reconnect with the version of myself I genuinely liked being.

That is when it hit me: rest is not a reward. It is part of the plan.

How Modern Life Hijacks Rest

Your brain was not designed for 147 browser tabs, literal or metaphorical.

Even when we are not working, we are working. Thinking. Scrolling. Toggling. Half-responding to a message while mentally prepping for tomorrow's meeting. We multitask our way through meals, errands, and even our downtime.

Modern life has turned recovery into background noise: always present, always postponed, never quite fully allowed.

We used to have natural rhythms. Day and night. Work and home. Off and on. But now, everything blends together. Email reminders at 9 p.m. Meetings creep into lunch. Social media serves us curated reminders of how much more productive, balanced, or well-rested *other* people seem to somehow be while running marathons, raising kids, and launching side businesses with artisan kombucha brands.

We are surrounded by what researcher Brigid Schulte calls "time confetti" [15]. These are tiny, constant interruptions that shred our day into unsatisfying scraps. Each alert, each task switch, each "let me just check one thing" steals a little cognitive energy and leaves us wondering why we are so tired even though we never did anything "that hard."

And the kicker? We call this normal.

We glorify the grind. We treat burnout like a personality trait. We wear sleep deprivation like a resume booster. The digital age has trained us to value availability over vitality, to be responsive, not rested.

Even when we try to rest, it is often passive. Phone in hand. TV on in the background.

Notifications buzzing like a mosquito we have learned to ignore, until we cannot anymore.

Here is the truth no app can fully fix: You cannot recharge in an environment designed to drain you. And unless we start reclaiming rest actively, intentionally, unapologetically, we will keep mistaking stimulation for energy, and exhaustion for just "how life is now."

Rest Is Not a Sin (Even If It Feels Like One)

We live in a world that does not respect rest. But it is not just your job or your calendar that makes it hard; it is the culture we grew up in.

Many of us were raised with some version of a work ethic where productivity was viewed as moral, and rest was indulgent, so we learned to stay busy. To keep moving. To always be useful. Even rest had to be earned.

Add to that the messaging we receive today, the Instagram reels that show 5 a.m. wakeups and bulletproof routines, the LinkedIn posts about side hustles before sunrise, and it is easy to feel like slowing down is some kind of failure. As if taking a nap makes you less ambitious. As if skipping a networking event means you do not care about your future.

We glorify busyness. We praise those who grind. We reward those who work late, answer emails on weekends, and say yes to every project. If someone posts a selfie from a hammock on a Tuesday, they better have a good excuse. Or at least a caption that says, "Do not worry, I am still checking my email."

At home, it is no easier. The dishwasher's full. The laundry is never done. The garage still has not cleaned itself. *I asked Alexa. She was completely useless.* There is a guilt that creeps in when you rest while someone else is working, even if no one has actually said anything. You internalize it. You negotiate with it. You try to "make up for it" later.

And that guilt? It is a byproduct of burnout culture, not a sign that rest is wrong.

The truth is that rest does not make you weak. It makes you *wise enough to listen.*

You do not have to prove your worth through exhaustion. You do not have to apologize for being human.

You just have to stop measuring your value in how many things you get done before you allow yourself to pause.

Rest as Strategy: High Performers Know the Secret

Here's a secret: the most successful people you admire? They do not grind 24/7. They incorporate recovery as part of their work because it is.

Take elite athletes. Every professional training plan, from Olympic sprinters to Ironman triathletes, includes structured recovery blocks. Not optional. Not a bonus. Required. Why? Because the real gains come when the body rests. It is not the reps that build strength, it is the repair.

Steph Curry? He tracks his sleep stats like most of us track our screen time. LeBron James averages 12 hours of sleep a night during training season. Simone Biles has spoken openly about stepping away from competition, not because of injury, but to protect her *mental health.* That is leadership.

And it is not just athletes.

Writers like Cal Newport, author of *Deep Work*, schedule "shutdown rituals" at the end of the workday; a hard stop to protect the evening as sacred recharge time. CEOs like Arianna Huffington build entire companies, like Thrive Global, around the science of sustainable performance, grounded in sleep, mindfulness, and boundaries.

And if you have read *Rest* by Alex Pang, you will know his big thesis: rest is not the opposite of work, it is the partner of it. The world's most creative minds, from Darwin to Maya Angelou, structured their days with long walks, naps, unstructured thinking time, and strict boundaries around their attention.

These people are not lazy. They are intentional. They have realized that pushing harder is not always the answer. Sometimes the most brilliant move is to stop.

If they can do it, with the pressure of world titles, boardrooms, and public scrutiny, we can too.

AI, Perfectionism, and the Rest Trap

AI was supposed to make life easier. Faster. Smoother. And in many ways, it has.

But it has also fed my perfectionism. Because now, with tools that let me edit endlessly, tweak fonts in seconds, or generate fifty headline variations instantly, I spend more time than ever trying to get it just right.

We think that advanced tools will save us time, and sometimes they do. But more often, they just raise the bar. Now that I can edit and optimize constantly, shouldn't I?

This creates a new kind of pressure, a digital hustle that never ends. And when you are stuck in that loop, rest feels like falling behind.

But the truth is, we were never meant to run at the pace of a machine. We are not robots. We do not have infinite processing power. We are beautifully, frustratingly human, and that means we need time to power down.

What Rest Looks Like (When It Actually Works)

Let me be clear: rest is not just about sleep, though sleep is critical.

Proper rest is about creating space to recharge, physically, mentally, and emotionally.

For me, that means:

- Morning meditation with my first cup of coffee before the kids are up, before notifications start, before the world wakes up.
- Reading a physical book before bed, something that slows me down and signals that the day is done.
- Stretching or yoga on training days, active recovery that calms the nervous system and clears my head.
- No phones at dinner, but honest conversations and tangible presence.
- Putting the phone away at night so my brain can decelerate without blue light and dopamine loops.

These are not massive lifestyle overhauls. They are small rituals that protect my energy and my peace of mind.

The Recovery Triangle: Your Energy Needs a Three-Part Strategy

Just like an endurance athlete cannot train endlessly without breaking down, your body and brain need more than willpower to recover. They need a system. You can call it the Recovery Triangle, and when it is in balance, you operate at your best: mentally clear, emotionally stable, and physically resilient.

It is built on three interconnected pillars: Rest, Fuel, and Movement. Missing one causes your system to strain. But when all three are present and supported, you feel like yourself again.

Rest

This is the essential, non-negotiable layer. You need 7–9 hours of sleep, ideally consistent, not just catching up on weekends. You also need intentional breaks during the day, even just a 5- to 10-minute pause to reset your attention and breathing. Rest does not just mean doing nothing. It means creating space for recovery; physically, mentally, and emotionally. Whether it is a nap, a nature walk, or a quiet cup of coffee, your nervous system needs periods of low stimulation to stay healthy.

Fuel

This is about what you put into your body and how it supports your brain. A steady flow of nutrient-rich meals stabilizes energy, mood, and focus. Hydration matters, too; even mild dehydration can impair cognition and mood. And yes, caffeine can be helpful in moderation. But if your second cup is just masking exhaustion, it is a short-term fix for a long-term problem. Food is not just fuel for the body; it is chemistry for the mind.

Movement

This is not about performance. It is about restoration. Daily movement, even gentle, improves circulation, boosts cognitive function, and helps flush stress hormones like cortisol. Think stretching, yoga, walking, not just workouts that leave you drenched in sweat. Even a quick 10-minute walk between meetings can reset your stress response and sharpen your attention. The goal is not to exhaust yourself; it is to reconnect your body and mind.

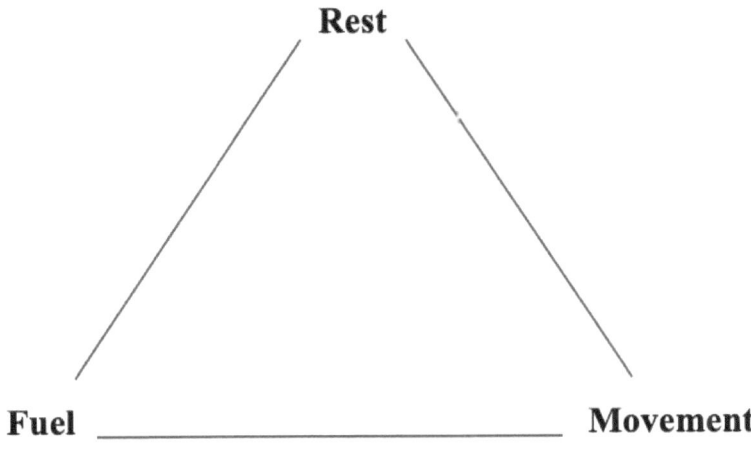

The Recovery Triangle

When these three elements are in balance, you don't just feel "not tired", you feel restored.

You can show up with confidence. Think clearly. Be creative. Handle pressure. And perhaps most importantly, you can enjoy life again.

Active vs. Passive Rest: Not All Downtime Is Equal

Let's clarify something: not all rest involves lying on the couch doing nothing, even though that is entirely valid and sometimes necessary. Rest comes in many forms, and one of the most impactful shifts we can make is to stop thinking of it as simply the absence of effort.

There are two main types of rest: passive and active. You need both. But depending on your stress levels, personality, and current pace of life, one might benefit you more than the other.

Passive rest is what most people imagine when they hear "recovery":

- Sleep
- Naps
- Lying down
- Zoning out on the couch
- Floating in a pool. *Highly recommended, by the way*

It is deeply restorative, especially when you are physically or emotionally exhausted. However, it is not the only way to recharge.

Active rest involves engaging your body or mind gently. It focuses on slowing down rather than stopping altogether. Think of it as recovery in motion.

Examples:

- A slow walk without headphones, just you and the sidewalk.
- Stretching or light yoga without any performance goal.
- Gardening, doodling, journaling.
- Listening to music while folding laundry at a relaxed pace.
- Making soup from scratch while humming badly to yourself.

These moments may seem ordinary, but they foster mental spacious-ness, giving your brain a chance to downshift, your nervous system to calm, and your sense of self to reemerge.

Sometimes we avoid passive rest because we associate stillness with laziness. And sometimes we skip active rest because we feel like it "does not count." But the truth is, both are essential. The key is intentionality; choosing rest rather than falling into collapse.

Here is the test: if your recovery time leaves you feeling more grounded, less frazzled, and more you afterward... it counts.

What Science Says About Rest

- Sleep supports memory consolidation, immune function, and emotional regulation. Lack of it increases the risk of anxiety, depression, and burnout.
- Breaks throughout the day help reset attention and reduce mental fatigue. Even five-minute "microbreaks" can restore energy and focus.
- Active recovery, such as walking, yoga, or light stretching, boosts

circulation, lowers cortisol: *bye-bye stress!*, releases endorphins to improve mood, and helps the brain switch out of fight-or-flight mode.

· Deep rest, the kind that involves genuine quiet, alone time, and stillness, helps your body switch into its natural 'recharge mode,' the system that restores health and energy over the long term.

If you are feeling exhausted, it is not because you are broken. It is because you are human, and all humans require recovery.

SIDEBAR: Warning Signs You Are Running On Empty (Even If You Are Still Getting Stuff Done)

Your body and mind will whisper before they scream. Here is what to watch out for:

· You wake up tired, even after 7+ hours of sleep.
· You get irritated by small things: traffic, typos, your own face in the mirror.
· You stare at a task for 15 minutes and forget what you were doing.
· You have "weekend crash" syndrome; you collapse once the adrenaline wears off.
· You over-caffeinate to *feel normal.*
· You keep telling people *and yourself*: "It is just a busy season," and that season never ends.
· You forget joyful things. Not because you are unhappy, but because you are too tired to enjoy them.

If any of this sounds familiar, it is not a sign of weakness. It is a sign of overdrive. Your system is not broken; it is *overdue* for rest.

You Do Not Have to Earn This

You don't need permission to rest *but I'm happy to give it if you want.*
    You do not have to hit a quota first.
    You do not need to justify taking care of yourself.

Rest is *not* a reward. It is a necessity.

You just need to notice when your body or mind is asking for a break and listen. This is not weakness; it is wisdom. The more you practice this, the more you realize that rest is not the opposite of productivity; it is what makes productivity sustainable.

Because the goal is not to do more, the goal is to feel alive while doing what matters.

The world will not always give you permission to rest. But that does not mean you have to wait for it.

So here is your permission slip:
    You do not have to be exhausted to take a nap.
    You do not have to justify a walk in the sun.
    You do not have to explain why you are closing your laptop at 5 p.m. instead of 9 p.m.

You do not have to earn rest. You are not a machine. And you were not put on this earth to get through your task list. Real productivity starts with honoring your limits.
    Rest is not a fallback plan. It is the foundation. And every time you choose it, you are not falling behind; you are showing up better for the things that actually matter.

* * *

Clarity Check-in

Think about your current rest habits:

- When was the last time you truly felt restored: mentally, emotionally, or physically?
- What stops you from resting when you need to? Is it guilt, pressure, or fear of falling behind?
- How do you usually try to rest and does it actually recharge you, or just numb you?
- What signals does your body give you when it is running low on energy?
- What small boundary or ritual could you add this week to protect your energy?

Write a *"Rest Permission Slip"* for yourself. Keep it visible. Use it when guilt shows up.

Experiment: Schedule a 20-minute *active rest* break daily: walk, stretch, quiet teatime. Treat it like a meeting you cannot cancel. At the end of the week, reflect: did those breaks shift your energy, mood, or focus?

Want to go deeper? In the Clarity Check-In section at the end of the book, you'll find a complete Energy Audit exercise to help you map your daily energy highs and lows.

* * *

6

The New Rules of Getting Things Done

Productivity should serve your well-being, not the other way around.

Hi. My name is Chris, and I have too many productivity tools.

It is a Monday, and I am already behind schedule. According to my five productivity systems, I should have been halfway through writing a blog post, halfway through planning my week, and ideally halfway through my second cup of coffee. Instead, I was halfway through an anxiety spiral.

My phone reminder was telling me to "prioritize what matters." My productivity app was reminding me to "brainstorm three podcast topics." My email app wanted me to respond to a calendar invite from last Thursday. And my personal calendar had just cheerfully notified me that "Chris is supposed to be at a dentist appointment in 15 minutes." *Thanks, traitor.*

I sat frozen at my desk, with a flashing instant message from someone asking if I had "a few minutes to talk about something quick." I did not. I never do. But I said yes anyway.

And that is when it hit me: I was not managing my time; I was drowning in it.

Productivity, somehow, had become its own full-time job. I had built an entire architecture of tools and routines designed to make life easier, and now they were the very thing overwhelming me.

So I stepped back. Asked better questions. And started rewriting the rules, not for my output, but for my sanity.

When "Productivity" Makes You Less Productive

You are not lazy. You are just drowning in optimization.

The old model of productivity gave us a clear blueprint:

- Optimize every minute
- Track every task
- Use every tool
- Stay "on" all the time

And it *worked*, for a time... Until it didn't.

We felt in control. There is something deeply satisfying about checking boxes, color-coding priorities, or syncing three calendars before 8 a.m. But at some point, that system tips. Instead of feeling clear and capable, you start feeling like you're falling behind. Instead of tools creating space, they create noise.

Eventually, you spend more time tending to your system than actually doing the work. You rearrange lists, tweak task labels, and set up automation that does not quite automate what you need. You bounce between tabs like you are solving a digital escape room and still forget your meeting is in five minutes.

It is productivity as performance. The illusion of efficiency without the experience of progress.

And worse? It chips away at your self-trust.

Because the more overwhelmed you feel while doing everything "right," the more you start to wonder:
   "Is it me? Am I broken?" *You are not.*

This is not a failure of discipline; it is a failure of design. You were never meant to juggle five systems, twenty tasks, and infinite notifications while staying calm, focused, and creative. That is not productivity. That is cognitive burnout in a productivity costume.

It is time to stop optimizing for speed and start designing for sanity.

What We Got Wrong About Productivity

We treated output like identity. And it broke us.
   Somewhere along the way, productivity stopped being about getting things done and started being about proving our worth.

We inherited the idea, whether consciously or not, that busyness equals value. That a full calendar means you are important. That exhaustion is a sign of dedication. That rest must be earned, and only after you have nearly burned out, as if fatigue were a badge of worth.

This mindset did not come out of nowhere. It is deeply rooted in industrial era thinking, where humans were expected to operate like machines; efficient, obedient, and always "on." Clock in, produce, repeat.

And in the digital age? We made it worse.

Now we can work from anywhere, at any time, which is precisely what we do. We blur the line between work and life until there is no line left. Productivity influencers showcase their optimized routines as lifestyle trophies. Our morning coffee is not just coffee; it is "part of my 5 a.m. habit stack." Every walk must be a podcast. Every hobby must become a side hustle. Every moment of rest is suspiciously close to laziness.

Most of these systems were designed for idealized, hyper-focused, neurotypical brains, not for real people navigating stress, trauma, caregiving, executive dysfunction, or chronic distraction.

So we start to internalize the guilt.
  "If I did not finish the list, I must have failed."
  "If I am tired, I must not be trying hard enough."
  "If I cannot keep up, I must not be good enough."

But what if we were never meant to "keep up" with a machine-paced world?

What if true productivity is not about more output, but about better alignment between your energy, attention, and priorities?

When we let go of the myth that busy equals valuable, we can finally build systems that work with our brains instead of against them.

Too Many Tools, Not Enough Clarity

At one point, I was using three different calendars: Microsoft Outlook for work, Apple for personal, and a paper planner for the "big picture." Add in Todoist, a productivity app, for tasks, ChatGPT for brainstorming, and Microsoft CoPilot for summarizing notes, and I was not just managing projects, I was managing my own digital chaos.

It became clear to me that more tools do not equal more clarity.

They often generate *more noise.* More friction. More anxiety that something has been forgotten, buried in the digital graveyard of old task lists and broken integrations.

The key shift? Consolidation and intentionality.

I started streamlining. I picked my primary calendar. I settled on one productivity app as my one source of truth for daily tasks. I stopped using three AI apps for the same thing and committed to the one that worked best for me. I gave myself permission *not to use every shiny new tool* that is labeled the next big thing.

Because the tool is not the strategy. The system is.

Designing Systems That Support Your Mind

Let's be clear: productivity tools are not the enemy. They are incredibly helpful *when they serve your actual needs.* Not your aspirational self. Not someone else's system you saw on YouTube.

Your actual mind, energy, and life.

The problem? Most systems are built for your ideal self, not your real one.

You know the version of you that wakes up early, drinks water before coffee, flows smoothly from deep work to errands with grace, and never gets sidetracked by snack cravings, cat video rabbit holes, or existential dread? Yeah. That person does not exist.

So the first step in building a better system is not downloading another app; it is accepting your brain as it is. Not how you wish it were. Not how someone on YouTube says it should work. Your actual brain, energy, attention span, and life demands.

That might mean:

- You need reminders to rest, not just reminders to hustle
- You function better with visual boards than task lists
- You overestimate what you can do in a day and underestimate how much stress that adds
- You need a system that still works on your worst days, not just your best

If you are neurodivergent, ADHD (attention deficit hyperactivity dis-order), autistic, or highly sensitive; the stakes are even higher. Most "one-size-fits-all" productivity advice was not built for your wiring. Systems that look perfect on paper often collapse under real-life energy swings, sensory overload, or attention fatigue. That is not your failure. It is a mismatch between your mind and your tools.

One shift that helped me? Stop mimicking other people's dashboards and start designing for clarity. My system does not have to be beautiful. It has to be breathable. Something I can use even when I am tired, distracted, or not feeling 100%. Here is what has worked for me and might work for you:

Weekly Planning Ritual

Every Friday afternoon, I block time to:

- Review what I accomplished
- Reflect on what drained or energized me
- Prioritize 3–5 core outcomes for the next week
- Declutter lingering tasks that no longer matter

It is simple. It is tactile sometimes. But it clears the mental fog and helps me start each week with intention, not stress.

Task Batching by Theme

Instead of jumping between 10 different tasks in one day, I assign "themes" to each day:

- Mondays: blog content and outlines
- Tuesdays: podcast scripting and editing
- Wednesdays: recording day

- Thursdays: newsletter and website updates
- Fridays: planning, admin, and catch-up

It is not rigid. Life happens, but it gives structure to the chaos. And most importantly, it keeps me from switching gears every 15 minutes, which is the fastest way to burn out your mind.

Ask yourself:

- What part of my current system actually works?
- What part feels heavy or guilt-inducing?
- What would it look like to build a system that supports how I feel, not just what I do?

Because the goal is not to become more productive. The goal is to become more you, with less stress, more intention, and a system that supports your life, not the illusion of someone else's.

AI Without Overwhelm

Let's talk about AI.

It is incredible. ChatGPT has assisted me in brainstorming blog post topics, outlining chapters for this book, planning podcast scripts, and even fixing website issues. CoPilot helps summarize articles and generate rough drafts faster than I could do on my own.

But here is the trap: AI can make you feel like you should *never stop improving.*

You can generate a dozen variations of everything. You can revise, polish, and perfect endlessly. I have fallen down that dark hole more than once, tweaking fonts, rewriting headlines, trying different visual layouts, only to end up with something 2% better after two extra hours.

That is not productivity. That is AI-powered perfectionism.

The AI Optimization Trap

AI is amazing. I am not here to argue otherwise. It has helped me brainstorm blog topics, summarize articles, outline this book, and debug the occasional website tantrum. It is fast, it is efficient, and it never judges me for starting a sentence five different ways before deciding on the first one.

But there is a sneaky side effect of having these tools at our fingertips: With AI, "done" has stopped feeling final.

Because now, we can generate ten more options. Rephrase it again. Try a different layout. Use a different format. Ask for more variations. Soften the tone. Tighten the wording. Add more polish.

No wonder the task we expected AI to cut in half ends up taking twice as long... and it somehow feels less satisfying when it's finished.

Welcome to AI-powered perfectionism, where good enough never truly feels good enough, and the urge to do more becomes a burden instead of an advantage.

This is a new form of "scope creep," but it happens inside our own minds. Without a clear stopping point, a sense of finality, or permission to say: "This is fine. I am done."

It is productivity inflation, and it is exhausting.

So, how do you avoid it?

By giving yourself creative boundaries. Set a limit on drafts, rounds, or prompts. Use AI like a sous-chef, not a taskmaster. Trust your first instincts more often. And remember, you are not obligated to optimize everything just because the tools exist.

Sometimes, progress is about finishing the task, not perfecting it endlessly.

AI can help you move faster, but only if you know where you are heading and when to stop.

The New Rules of Getting Things Done (In the Age of AI)

Let's face it, the old productivity playbook does not hold up in the age of AI, constant updates, and mental overload. So here are three new rules to help you stay productive without losing your mind, or your weekend.

Rule 1: Be Self-Aware Before You Plan

Before you open any productivity app or write down your top three priorities, take a moment to pause. Ask yourself:

- How am I feeling today?
- Do I have the energy to focus, or do I need a reset first?
- What kind of work would feel *good* to start with, not just what is "urgent"?

You can have the most polished task system on Earth, but if your brain is foggy or your emotional bandwidth is depleted, nothing's going to click.

This rule is about meeting yourself where you are, not where you wish you were. Because pushing through burnout with a color-coded list is like trying to fix a dying plant with a new pot. It is the wrong problem for that solution.

Rule 2: Think Before You Prompt

AI is your co-pilot, not your brain.

AI can generate ideas, summarize research, and refine your voice. But it cannot do your thinking for you; if you rely on it too much, you will start losing touch with your own creative instincts.

Before asking ChatGPT or CoPilot for help, ask yourself:

- What do I already know?
- What do I actually want this to say or do?
- What part feels unclear or blocked?

The more precise your input, the better the output. It also reduces the chance of over-revising or losing your creative connection.

Pro tip: Write one rough draft before prompting AI. Compare it with your AI-generated version, then combine the best ideas. This keeps your voice in the final piece instead of letting a polished AI response take over.

Rule 3: Do not Drown in Options

More drafts will not fix a shaky decision.

AI can give you 10 versions of anything. That does not mean you need to read them all.

Set limits:

- One round of brainstorming
- Two passes of revision
- A timer for tweaks: a maximum of 20 minutes, then ship it

Having more choices does not guarantee better results; they simply drain your decision-making energy. Think of your brain like a web browser: too many tabs open, and everything starts to lag. Sometimes, "good enough" is not a cop-out; it is a power move. The goal is not to perfect things forever. It is to move forward with clarity and confidence.

AI is a powerful tool. But it requires human boundaries. Your voice still matters. Your energy still matters.

You still matter.

Your Brain Is Not a Cloud Platform

You are not built for 24/7 uptime and automatic updates. Stop acting like it.

Somewhere along the way, we started treating our brains like software.

We expect ourselves to be:

- Always available
- Seamlessly integrated

- Fast, efficient, and infinitely scalable
- Capable of syncing across multiple roles, devices, and platforms without ever glitching

But the reality is, your brain is not a cloud-based app.

It does not run on clean code. It runs on blood sugar, sleep, stress hormones, and snack breaks.

It has limits. Fluctuations. Mood swings. It does not respond well to being pushed past those limits in pursuit of more optimization.

You do not "update" your mind by adding more plugins. You restore it through pauses.

Through disconnection. Through human things, like walking without a phone or thinking without a prompt.

Productivity culture often portrays success as frictionless, as if the goal is to become more like the tech we use. The truth is that friction is human.

You need downtime, glitches, space to reboot, and time to not know the answer right away.

You do not need to run your life like a startup dashboard. You need to run it like a living system, one that responds to seasons, energy, emotions, and rhythm.

So give yourself permission to work like a human, which means being imperfect, adaptable, and sometimes, gloriously offline.

When to Go Analog (and Why It Helps)

As digital as my life is, I still go analog, non-digital tools like paper and pen, when it matters. There is something oddly grounding about writing things down by hand, even if my handwriting looks like a confused squirrel tried cursive. When I am overwhelmed, I do not open a new app. I grab a journal. When I am trying to think clearly about a project, I sketch it out on paper. When I am juggling too many ideas, I brain-dump in my journal as if I am venting to a therapist who never interrupts.

Is it slower? Yes.

Does my hand cramp? Also yes.

But something shifts when you write things down. Your thoughts untangle. Your pace slows. Your *mind exhales.* If you have never tried pen and paper after a week of screen fatigue, give it a shot. You might be surprised what your brain's been trying to tell you.

Analog is not nostalgic; it is neurological.

Studies show that writing by hand activates different areas of the brain than typing. It slows your thinking just enough to clarify it. It improves memory, supports emotional processing, and reduces mental overload. That pause between thought and ink creates breathing room.

So if you have hit digital fatigue and your brain feels like it is buffering, try going analog. Give your nervous system something it can actually hold.

Balancing Effectiveness and Mental Health

At the end of the day, getting things done should not be the *goal*. It should be the byproduct of clarity, alignment, and energy.

We chase productivity like it is a finish line that keeps moving because, in many ways, it is. There is always another tab to open. Another tool to try. Another version to polish. Another thing to squeeze in before you can "call it a day."

But here is the twist: "done" is not a checklist. It is a state of mind.

It's recognizing that you did what mattered; not everything, not perfectly, but enough for today. It is stepping away without guilt. It is letting things remain unfinished when your body says it is time to rest. It is trusting that what is left undone tonight can still be okay tomorrow.

Sometimes, the most productive thing you can do is stop working. Sometimes, the most responsible choice is to say, "not now." Sometimes, the best work happens after you step away from the screen and touch grass. Or toast. Or silence.

"Done" does not mean depleted.

It means clear. It means aligned. It means you are still here, and you are still human, and that matters more than any task ever will.

So go ahead. Close the tab. Finish the day. Let yourself be done.

BONUS SIDEBAR: Signs Your System Is Not Serving You
    You might need a system reset if:

- You have tasks listed in five places and do not trust any of them
- You feel guilty looking at your own to-do list
- You spend more time organizing your tasks than doing them
- You constantly rewrite the same task every week
- You never feel "done," even when the day is over

Simplify. Consolidate. And make peace with not catching *every ball*. Just the ones that matter.

***

Clarity Check-in

Getting things done is not about checking more boxes, it is about checking in with yourself.

Let your tools serve you, not run you. Let your calendar support your mind, not hijack your time. And remember: you are still the point of the process.

Ask yourself:

- What is one tool, routine, or mindset you've been using that adds more stress than clarity?
- When you open your calendar or task list, do you feel supported or overwhelmed?
- Can you simplify it? Swap it out? Let it go?
- If you rebuilt your system from scratch, what would you *actually* keep?

Now, write down one "new rule" you want to live by. Not to do more, but to do what matters.

Stuck on where to begin? Don't worry, I have got you covered. In the Clarity Check-In section at the back, you will find all the exercises gathered in one place, plus a New Rules Starter Template to help you sketch out your own guideposts.

Experiment: For one week, limit yourself to 3 meaningful tasks per day. At the end of each day, track how you feel: more in control, less? More energized, less? By the end of the week, notice whether doing *less* actually helped you achieve *more of what matters.*

* * *

In the end, productivity systems and AI tools are only part of the story. What matters most is remembering that you are not a machine. You will have limits, quirks, and flaws. And that's not a weakness. It's the starting point for staying fully human in a machine-shaped world.

III

Part III: Staying Human in a Machine World

"He who has a why to live can bear almost any how." —
Friedrich Nietzsche

*Part II explored redesigning life and work to protect energy
and sanity. But even the best system fails if we keep
measuring against impossible standards. Part III is about
reclaiming what technology can't reproduce: your humanity.
Not the polished or optimized self, but the imperfect,
creative, emotional being who still shows up, stumbles,
connects, and makes life meaningful.*

7

Imperfection Is the Point

You are not a machine, and that is your strength.

We live in a world that practically worships polish. Clean interfaces. Flawless execution. High-performing teams and perfectly optimized routines. The irony? Most of us are barely holding it together under that glossy surface.

We scroll through curated feeds, listen to perfectly mixed audio, and read blog posts that sound like they were written by a team of editors. And somewhere in the back of our minds, a quiet voice whispers:

"If it is not flawless, do not share it."

And it is no wonder. We are living alongside tools, especially AI, that can spin out polished writing, strategic plans, recipes, lesson plans, ad copy, poems, and press releases in seconds. Seconds! And while that can be empowering, it can also leave us wondering:

"If the machines are perfect, do I need to be too?"

Short answer: no. Long answer: absolutely not, and here is why.

The Perfectionism Trap

Perfectionism does not usually show up waving a big red flag. It shows up quietly, disguised as "being thorough" or "having high standards." It tells you that you are just being responsible, that you are protecting your reputation, your credibility, your craft.

But underneath, it is rarely about excellence. It is about fear.

Fear of looking foolish.

Fear of being criticized.

Fear of confirming that little voice in your head that says you are not good enough.

Perfectionism is a moving finish line. No matter how much you tweak, there is always something else to improve. And in an AI-saturated world, that finish line can now stretch into infinity. The tools make it possible to revise endlessly, allowing for new variations, new angles, better formatting, and sharper grammar. And every improvement makes you wonder if there is an even better version waiting one prompt away.

Here is a quick gut check: you might be in the perfectionism trap if...

- You have spent more time formatting a document than writing it.
- You have delayed sending something because "it needs just one more pass."
- You have re-recorded audio or video multiple times for reasons no one else would notice.

- You feel anxious after sharing something because you are still thinking about what you could have changed.

The trap is not that you care about doing good work. Caring is great. The trap is believing that perfect work exists, and that it is the only kind that is safe to share.

The truth is, perfectionism does not protect you from criticism. It just delays connection. It pushes your ideas, your voice, your presence further down the road while someone else, flaws and all, is showing up and making an impact right now.

The Myth of Readiness

"I'll publish it when it is ready."

"I'll launch when I feel more confident."

"I'll share my story once I have figured it all out."

Readiness sounds noble. Responsible. Like you are being patient and thoughtful. But most of the time? It is a mirage and perfectionism in disguise.

Readiness whispers that there is a moment in the future when I will feel completely prepared. One more draft, one more late-night tweak, and one more round of feedback will finally give you the certainty you need. But that moment rarely, if ever, arrives.

Some of the most impactful things in history were shared before they were "ready."

- The first iPhone launched with no App Store.
- J.K. Rowling was still tweaking Harry Potter ideas between books.
- Countless start-ups have released beta versions riddled with bugs, only to improve them significantly once they were in the hands of real users.

Creativity lives in the messy middle, not the pristine finish line. The messy middle is where your work evolves in response to real people, honest feedback, and genuine connection. If you wait until you feel ready, you skip the stage that teaches you the most.

When I think back to the work that resonated most with people, including blog posts, podcast episodes, and presentations I have given, almost none of them were shared from a place of total confidence. They were shared in moments of doubt, when part of me wanted to hit delete instead of publish. And those are the same pieces that people email me about months later, saying, "This was exactly what I needed to hear."

The truth? You do not need to be ready. You need to be real.

Because readiness is not a prerequisite for impact. Honesty is. Vulnerability is. Showing up is.

The Illusion of "Flawless"

AI is dazzling. That is part of the problem. It does not stutter. It does not second-guess. It does not forget why it walked into the kitchen. It can produce eloquent paragraphs faster than most of us can find our reading glasses.

And it is easy to fall into the trap of measuring yourself against that kind of precision. After all, AI does not need breaks. It does not worry about impostor syndrome. It does not spill coffee on its laptop or accidentally send an email to the wrong person. *Shout out to everyone who has ever done that. I know a guy who hit reply-all to an email that included the CEO, who responded and chewed him out. Ouch.*

But you know what else AI does not have? Empathy. Nuance. Lived experience. Intuition. Humor that does not come off like a dad joke generator.'

Here is the real contrast:

AI Does...	Humans Do...
Generate perfect grammar every time	Write run-on sentences that accidentally become poetic
Respond instantly	Take a beat to read the room (or the mood)
Never need sleep	Create brilliance after a nap or a thought in the shower
Never forget a fact	Tell a story slightly wrong but make it funnier
Always "finish"	Stop halfway to go hug a kid, walk the dog, or make tea

Contrast Between AI and Human

The fact that you are human is not a liability. It is the source of your most valuable work. Because what people remember is not always the perfectly constructed thing, it is the thing that made them feel seen. And feeling seen does not come from polish. It comes from realness.

Case in Point: A Table Full of Flaws

Let me offer a very non-digital example. One of my favorite hobbies is woodworking. It is deeply satisfying to build something with your hands, especially in a world where we are so often just tapping glass screens.

But here is the truth about woodworking: nothing ever turns out perfect.

Ever. There are nicks and knots, uneven stains, and angles that are slightly off. And yet? That is what makes it mine.

A few years back, I built a table for our dining room. It is solid, it does its job, and from across the room it looks great. But if you get up close, you can see where I mismeasured a board by a 1/4 inch. You will spot the corner I over-sanded. None of that matters to my wife. She loves it. And honestly, so do I, because I made it. It has flaws, but those flaws make it mine. And it stands *literally*!

That table is not just furniture. It is proof of effort. Of process. Of patience. It is perfectly imperfect. And it's way more fulfilling than anything that could have come from a machine.

Another Case in Point: Cooking Without the Recipe Police

Woodworking is not the only place I have learned to embrace imperfection. Cooking has been just as humbling, and just as freeing.

When I cook, something almost always goes "wrong." The sauce does not thicken enough. The chicken's slightly overdone. I knock the flour on the floor. I dump an entire bottle of ground nutmeg into the blueberry cobbler... *For the record, I've been told to clarify that I am, in fact, an amazing chef and these mishaps say more about my clumsiness than my cooking.* If I were following the imaginary "Recipe Police" rulebook, these would be disasters.

But here is the twist: nine times out of ten, no one notices. Or they notice and prefer it that way. The unevenly chopped vegetables? "It looks rustic." The collapsed soufflé? "Oh, it is deconstructed."

Cooking taught me that the most memorable meals are not always the flawless ones. They are the ones made with care, shared with people you love, and served without apology for the little quirks along the way.

Just like the dining table I built, those imperfect meals carry a story. And that story, the effort, the improvisation, the personality, matters far more than whether the plating could win a Michelin star.

That is the beauty of embracing imperfection: it turns "mistakes" into flavor, in the kitchen and in life.

SIDEBAR: Signs You are Chasing Perfection and How to Flip the Script

Perfectionism wears many disguises. It does not always look like obsessively color-coded spreadsheets or 14 podcast draft edits. Sometimes, it is subtler. Sneakier. It shows up in the way we delay, overthink, and tie our worth to our work.

Here is a breakdown of everyday perfectionist habits and how to reframe them into progress-driven practices:

Perfectionist Habit	Progress Mindset
You delay starting until things feel " just right."	Start messy. Clarity comes through action.
You tweak, edit, or "just double-check" more than necessary.	Done is better than perfect. Publish the draft.
You struggle to delegate.	Collaboration builds trust and momentum.
You feel like your work represents your worth.	You are more than your output. Mistakes *do not* equal failure.
You avoid sharing things publicly until they are "polished."	Vulnerability connects. Raw work resonates.
You feel burned out but keep pushing anyway.	Rest is productive. Recovery fuels creativity.
You compare your process to others' final product.	Your journey is valid. Their polish is not your benchmark.

Perfectionist Habit versus Progress Mindset

Publicly Imperfect

When I launched my blog, The Mental Lens, and my podcast, Through the Mental Lens, I did what many creators do: I over-edited. I overthought. I delayed publishing because I was unsure if my message was "ready." I wondered if my voice was professional enough. If my layout looked sleek enough. If I was being vulnerable in the right way.

Then I published an episode about my own struggles with anxiety, something I almost did not share at all because it felt a little too real.

Guess what? It is still my most downloaded episode. I got messages from listeners saying, "This made me feel less alone." Not "this was perfectly produced." Not "you should be a voice actor."

Just... thank you.

And it is not just me. Think about Brené Brown's famous TEDx talk on vulnerability. The lighting was off. The staging was simple. She was not wearing a slick TV anchor outfit. But the message was raw, relatable, and human, and it became one of the most-watched TED talks of all time.

That is what imperfection can do. It builds bridges. It invites honesty. It makes space for others to say, "Hey, me too."

What AI Can (and Cannot) Do

As I keep stating, I use AI all the time. This book would not be what it is without my trusty co-pilot, ChatGPT. It helps me brainstorm. Organize. Refine. Rewrite. It is like having a super-smart assistant who never sleeps or judges your rough drafts.

But it is not me. It does not have my voice. My stories. My sarcasm. My messy humanity.

AI is great at producing ideas. But it is not great at *deciding* which idea feels most true. That is where you, the human, come in.

Here is a metaphor: if you are trying to build a campfire, AI can give you the wood, the kindling, even a flame. But only you know where to place the firepit, how high the flame should go, and how to keep it from burning the marshmallows.

Progress > Perfection

If I could tattoo one phrase on the collective consciousness of this generation, it would be:

"Progress over perfection."

Because perfection is not absolute, it is a moving target. It is the mirage that convinces you to keep going one more round, one more edit, one more tweak, and then maybe you will feel "ready."

Plot twist: I am not.

But progress? That is trackable. That is growth. That is you showing up today with 2% more courage than yesterday. That is learning a new skill, sharing a raw story, building a flawed but functional dining table, or taking a walk instead of finishing that 27th draft.

Progress is human. Perfection is a myth.

Psychologists like Dr. Kristin Neff have shown that perfectionism thrives on self-criticism, while progress thrives on self-compassion; like giving yourself the same kindness you would offer a friend. The data is precise: self-compassion fuels resilience, creativity, and motivation far better than an inner critic ever could.

Want a simple way to test it? Try the 80% Rule: when something feels "good enough" at about 80%, release it. Perfectionists often spend half their time chasing the final 5% of polish. Progress comes from finishing, not endless tinkering.

And in a world where AI can reword, redesign, and reframe everything instantly, the temptation to "perfect" is stronger than ever. But "better" is not always better. Sometimes the bravest thing you can do is to stop, release, and let imperfect work live in the world.

The Slow Path Is the Smart Path

We are not designed for the speed AI offers. Our bodies need rest. Our minds need space. Our creativity? It often needs boredom to spark.

Neuroscientists call this the default mode network; the brain state that activates when you are not actively focused on a task. It is the mental "idling" that happens during a walk, a shower, or staring out the window. This is when your brain quietly makes connections, solves problems, and generates fresh ideas.

Translation: Sometimes the best way to move forward is to step away.

I have learned this the hard way. More than once, I have tried to force a blog post into existence by hammering at my keyboard for hours, tweaking sentences until my eyes blurred. On one particularly stubborn writing day, I gave up and went for a 30-minute run. Somewhere between mile one and mile two, the exact sentence I had been struggling with popped into my head, fully formed. I was not even thinking about the post anymore, but the moment I stopped forcing it, my brain delivered.

Slowing down is not laziness. It is intelligence, especially when you are working on something meaningful. If you are building something that matters, whether a business, a book, or a relationship, the slow route is the sustainable one.

The irony? In a culture obsessed with speed, slowing down often puts you ahead. Because while everyone else is busy pushing harder, you are making space for clarity, energy, and ideas worth sharing.

Community: Rest We Share

There is another form of rest we rarely name: the kind that comes from being with others. When we think of recovery, we picture sleep, solitude, or meditation. But human biology is wired for connection, and science keeps proving it.

Psychologist Julianne Holt-Lunstad reviewed more than 300,000 participants and found that people with strong social ties had a 50% lower risk of early death than those who were isolated [17]. Another long-term study showed that meaningful social connections cut the odds of developing depression by nearly a quarter [18].

And perhaps the clearest picture comes from the Harvard Study of Adult Development. For over eight decades, researchers have followed people across their lifespans. The strongest predictor of health and happiness wasn't cholesterol levels, income, or even career success, it was the quality of their relationships [16].

That doesn't mean you need dozens of friends or a massive network. It means the small, steady circles matter most. The coworker chat that makes you laugh after a brutal meeting. The neighbor who waves every morning. The running group that shows up even when it rains. These tiny communities don't just brighten your day, they buffer your stress.

And the best part? Building that kind of community doesn't require grand gestures. It can be as simple as:

- Inviting a friend for a standing weekly coffee or walk.
- Joining a group built around something you already love like a book club, a class, a hobby.
- Being the one to send the first "thinking of you" text.
- Starting small rituals like Friday check-ins, monthly dinners, or even group memes that spark joy.

Community is not extra. It is not a luxury add-on to wellness. It's rest we share. The kind that reminds us we don't have to carry everything alone.

What AI Gets Wrong

The reason AI outputs feel so clean is that they do not come with the friction of being human. No procrastination. No fear of rejection. No over-caffeinated, under-rested second-guessing. Just structured language, sourced from billions of inputs, spat out in milliseconds.

That polish can be impressive and intimidating. If a machine can produce a flawless draft in seconds, it is tempting to think we should work faster, cleaner, and better, too. But here is the flaw in that thinking: flawless is not the same as valuable.

AI cannot do lived experience. It cannot embed the story about how you built a table that wobbled for the first month. It cannot replicate the mix of pride and embarrassment in serving a slightly overcooked meal to friends. It cannot understand why you paused mid-sentence in your podcast to take a breath because the topic hit too close to home.

It can simulate human voices, but it does not live a human life. And that matters, because the moments that connect us, the ones that make people lean in and think, "I get this person," are not usually the shiny, polished bits. They are the minor imperfections, the pauses, the unscripted moments.

If we start competing with AI on its terms of agility, accuracy, and flawlessness, we give up the very things that make us memorable. We trade our human fingerprints for machine polish. And that is not the game we want to play.

Our weirdness, our quirks, our contradictions; those are not bugs to fix. They are features worth protecting.

You are Not a Robot and That Is the Point

Robots do not get distracted when their kid walks into the room mid-meeting. They do not need mental health days. They do not overthink a text at 2 a.m., ugly cry during a Pixar movie, or laugh so hard they snort in the middle of a serious conversation.

You do. I do. That is not a weakness; that is the whole point.

In a world that is moving faster, optimizing harder, and outsourcing more to machines, the most valuable thing you can bring to the table is the thing no machine can imitate: your humanness.

That means:

- The slowness that lets you notice details others miss.

- The emotion that makes your work resonate.
- The nuance that comes from lived experience.
- The mistakes that teach you more than success ever could.
- The growth that comes from showing up anyway.

Your humanity is not something to smooth over. It is the reason your work matters. And in the long run, it is the one competitive advantage AI will never have.

So stop trying to work like a machine. Let yourself be a little rough around the edges. Keep the pauses. Keep the quirks. Keep the version that is real, not just polished.

Because when the world is flooded with flawless outputs, the flawed, human ones will stand out. And those are the ones people will remember.

* * *

Clarity Check-in

Ask yourself:

- What is one area of your life where you are holding yourself to an unrealistic standard?
- Where does perfectionism show up most often? Work, parenting, relationships, or even hobbies.
- What is the cost of chasing perfection? (lost energy, delayed

progress, missed joy)

- Whose approval are you actually trying to earn, and does it matter as much as you think?
- What would happen if you let go of that perfection, just a little, and gave yourself permission to show up as you are?

Experiment: This week, share or publish one piece of work "as is". No extra polishing, no endless tweaks. Then, notice how people respond compared to your usual process. More importantly, how do *you* feel? Lighter, anxious, relieved, or maybe even proud? Capture that response and remind yourself: imperfect still counts.

* * *

Imperfection, In Practice

As you close this chapter, remember: The things you build, your work, your relationships, your ideas, do not need to be flawless.

They need to be real. Useful. Meaningful. Human.

So go ahead. Share the blog post with a typo. Record the video in one take. Wear the shirt with the coffee stain on your video call. Publish the draft before it is perfect.

Because perfection is not what people remember.

People remember the table you built with your hands. The podcast where your voice shook a little. The story that made them feel less alone.

People remember you: flaws and all.

8

Emotions, Meaning, and the Stuff You Cannot Automate

Mental health is not a productivity hack; it is the foundation.

It was the weekend, and my oldest had just finished drawing me a picture, this time a freehand dragon, complete with a beautifully contrasting, colored background. Seriously. This kid has talent. Yes, I am biased.

"Do you like it, Dad?" he asked.

I did. I loved it. But here is the part I am not proud of: before answering, my brain instinctively thought, I should scan and post this. People would love it.

In less than a second, I had shifted from feeling the moment to wondering how to package it. From connection to content. From meaning to metrics.

That is the world we live in now. We are surrounded by tools that can capture, optimize, and share everything, but they can also distract us from truly engaging with our experiences.

And the more we try to engineer joy, purpose, or peace of mind, the more we forget the simple truth: you cannot automate what makes life matter.

Clarity, meaning, and mental health are not features to toggle on or off. They are not downloadable. They are human. And human means messy.

The Limits of the Upgrade

We love upgrades. Faster phones. Smarter watches. Apps that promise to help us sleep, breathe, focus, or eat better. And sometimes they work for a while.

But some upgrades do not really upgrade us at all.

Like the meditation app that reminds you of "daily streaks" until mindfulness feels like a scoreboard. Or the AI-generated greeting card that says all the right words but somehow still feels hollow. Or the fitness tracker that congratulates you for hitting 10,000 steps, even if most of them were pacing in your kitchen while stress-eating cookies.

These tools are not bad. They are just not the same as the things they are trying to replace. Tracking sleep is not the same as feeling rested. An auto-generated "thinking of you" message is not the same as hearing a friend's voice. A shared Google album is not the same as flipping through a photo book with someone you love.

Some parts of life just are not meant to be optimized. They are intended to be experienced; messily, slowly, and without a "share" button.

When Feelings Guide the Fork in the Road

Let's get personal for a minute.

There have been moments in my life where emotion almost steered me off a cliff. I have nearly quit big things, important things, simply because something did not go my way. I felt out of control, so I wanted to regain control the only way I knew how, by walking away. By letting the emotion take the wheel.

But then I learned the power of pause. Of breathing. Of stepping back and letting the wave pass.

Take The Mental Lens, for example. What started as a bit of research into side hustles quickly became something personal. Emotional, even. I was not looking to start a blog or podcast about mental health. But after battling anxiety, recognizing how many others were doing the same silently, and realizing how much bad advice was out there, I made the decision to jump in.

That choice was driven by emotion, but not in a reckless way. It was purposeful. Empathy-fueled. And it continues to be.

Another time, I was training for an Ironman I had been working toward for months. An injury flared up a few weeks before the event. My first instinct? Scrap the whole thing. Save face. Avoid the disappointment of a slower finish, if I could finish at all. But after a few days of frustration and self-pity, I realized my real goal was not a perfect time; it was proving to myself that I could see it through. I adjusted my training, showed up on race day, and crossed the finish line smiling.

Slower than planned, but prouder than if I had quit.

Emotions can push us toward the exit. But they can also pull us toward the things that matter most, if we are willing to pause, listen, and choose with intention.

What AI Will Never Feel

Let's state the obvious: AI can mimic emotion. It can respond empathetically. It can even help us process emotions by organizing thoughts, generating prompts, or simulating conversation.

But it does not feel, *at least not during the process of writing this book.*

It does not know heartbreak. It does not understand what it means to hold your child after a hard day. It does not get goosebumps when your favorite song hits that perfect note. And that is not a flaw.

It is the line.

I have tried to use AI for nuanced, emotional writing; letters, reflections, even parts of this book. And sometimes, the result feels flat. Not because the words are wrong, but because the *weight* is not there. There is no lived experience behind it. No shared struggle. No personal risk.

And that risk? That is what makes human emotion *real.* It is what makes meaning matter.

AI will never experience the everyday "human glitches" that give those emotions context, like:

- Forgetting why you walked into the kitchen twice.
- Crying over an episode of Bluey... and not being able to explain why.
- Stopping mid-run because the sunset looked too good to ignore.
- Re-reading an old text from someone you miss and feeling it all over again.
- Burning dinner because you were too busy laughing with a friend.

These are not inefficiencies. They are proof of life: *remember this when the house is a disaster and the kids' stuff is everywhere.* The things that break our rhythm are often the very things that make life worth living. And no amount of processing power can replicate that.

Empathy as a Superpower

Years ago, I made a decision: if I knew how hard anxiety and uncertainty could feel, I wanted to help others through it too. I did not need to become a therapist or a life coach. What I discovered instead was the role of a *sign-poster.*

So what does that mean? Imagine being in an unfamiliar airport or hospital. You do not need someone to walk you all the way to your gate or room. What you really need is a clear sign pointing you in the right direction. That is what signposting is: you are not the solution, but you help someone get closer to one.

At my company, this idea evolved into something much larger. I joined the Mental Health Champion program, a global peer-to-peer network of colleagues trained to offer confidential, empathetic support. The role is simple but powerful: listen without judgment, normalize the conversation, and point people toward the right resources, whether that is an Employee Assistance Program, a trusted toolkit, or just permission to take a mental health day. Sometimes, advocacy looks less like a dramatic rescue and more like a quiet sign that says, *"You are not alone, here is where you can find help."*

And here is the part I love: you do not need a certification to be that kind of ally. Anyone can put up signposts.

- It may involve keeping a folder of vetted mental health resources to share with a colleague or friend.
- It could be tacking a flyer with hotline numbers in the break room.
- Perhaps it involves creating a shared space for wellness tips, as seen with the IT specialist I read about, who started a "Wellness Resources" channel on Microsoft Teams. At first, only a few people noticed. But as others added their own suggestions, it grew into one of the most active spaces in the company, all because one person decided to put up a signpost.

That is the essence of this work: you do not drag people down the road; you make sure they know the road exists.

When I became a certified Mental Health Champion, I realized how many people are quietly struggling. And not just struggling, but struggling in complete isolation. Since then, countless coworkers have opened up to me. Not because I had the right answers, *I rarely did*, but because I listened. Because I had been there. Because I knew how it felt to show up at work smiling while anxiety gnawed at me inside.

And that is where empathy comes in. My kids remind me of this daily. They do not always need me to solve their problems. In fact, sometimes, *okay often*, they resist advice. What they do need is for me to sit with them in the frustration of a broken Lego tower or acknowledge how scary it feels to stand on the edge of the diving board. That act of *being with* is the heart of empathy.

I have felt it on the other side as well. In those moments when someone sat with me in my own mess, offering no fixes, just presence. That is when I realized: empathy is not about perfection or expertise. It is about humanity.

And that is why empathy is a superpower. It cannot be automated or coded. Genuine empathy lives in the messy, unpredictable space where two people meet exactly as they are, without a script, without rushing to tidy it up. It is not about solving. It is about seeing, hearing, and being present.

One of the simplest ways to turn empathy into practice is to change the questions we ask. "How are you?" often gets a reflexive "fine" or "busy." But swap it for something more specific, and you open the door to a real exchange.

Here are a few examples that have made a difference in my own life:

- Instead of: "How's it going?"
- Try: "What's felt heavy for you this week?"
- Instead of: "Did you have a good day?"
- Try: "What's one moment that stood out to you today?"

- Instead of silence when you see someone struggling:
- Try: "I don't want to fix this for you, but I want to hear it. What's been hardest?"

These aren't magic formulas, and they won't land perfectly every time. But they signal that you're willing to go beneath the surface; that you want to know more than the highlight reel.

And the data backs this up. According to **Mental Health America**, 71% of people turn to friends or family first when they're stressed [19]. That means every small choice we make to ask a deeper question can become part of someone's coping strategy.

When you ask questions like these, you're not just gathering information. You're sending a message: *"Your inner world matters to me."* That message alone can make someone feel lighter, even before a single solution appears.

Meaning > Metrics

There is a quiet lie baked into modern productivity: that metrics equal meaning. That if we hit all our deadlines, check all the boxes, and optimize every process, we will finally feel fulfilled.

Guess what? We will not.

My sense of purpose does not come from my inbox. It comes from being a dad. From showing my kids what it looks like to train for a race, fall short, and try again. From hugging them when they are upset and letting them see me do the same. It comes from helping someone take their first step towards therapy. From writing a post that helps someone feel seen.

Metrics can be useful. They can show us patterns, track progress, even help us make better decisions. But they are not the same thing as meaning.

Metrics can tell you:

- How many people opened your newsletter.
- How fast you ran that 5K.
- How many likes your post receive.

Meaning tells you:

- Who read your words and finally booked a therapy session.
- What the race taught you about resilience.
- Which post made someone feel less alone.

Numbers can measure reach, meaning measures impact. And the latter often happens quietly, without a single notification.

The truth is, automation can remove friction, but it can also remove connection. We delegate tasks and lose conversations. We automate workflows and forget the *people* behind them.

Efficiency without empathy is a hollow win.

The (Actual) Value of Mental Health

Somewhere along the line, mental health got marketed as a performance enhancer.

"Get better sleep so you can work harder."

"Meditate to increase focus."

"Exercise to improve productivity."

Now, do not get me wrong, I love a good boost from a hard workout. But mental health is not a strategy to *do* more. It is the foundation that lets us *be* more.

When your mental health is shaky, everything else is harder.

When it is strong, even the hard things feel more manageable.

Here is the difference:

Before - running on fumes:

- You wake up already tense, thinking about the unfinished tasks from yesterday.
- You power through meetings, but you can't recall what was actually decided.
- Every new email feels like an ambush.
- You hit the end of the day exhausted but weirdly unsatisfied.

After - working from a healthy baseline:

- You start the morning feeling grounded, not frantic.
- You can respond thoughtfully instead of reacting impulsively.
- You make better decisions because your brain is not stuck in fight-

or-flight mode.

· You end the day tired but content, with enough energy left to enjoy your life outside work.

Mental health is not a side quest. It is the main quest. It is the difference between feeling like you are constantly trying to "catch up" and feeling like you are actually in control of your life.

And I wish more people understood that healing is not *linear*. By that, I mean it does not move in a straight line. You do not wake up one day, decide to work on your mental health, and then feel steadily better each week until you are "done." Real healing looks more like a messy graph: progress one week, setbacks the next, and slow, uneven gains over time. Some days you feel like you are moving backward, even though the overall trend is forward.

My own mental health journey has been anything but clean. Even after seeking help, I did not feel relief overnight. It took work. Awareness. Trial and error. And plenty of days where I thought, *Shouldn't I be past this by now?*

But every step forward was worth it.

If there is one thing I would tell someone who feels like emotions are getting in the way of their productivity, it is this:

Emotions are not the obstacle. They are the indicator.

They tell us when something is wrong. When something matters. When something is healing.

You Cannot Automate Awe

I love woodworking. It is my imperfect, messy, human hobby. I build things with my hands. Sometimes they are beautiful. Sometimes they are crooked. But they are mine.

I could imagine a future where robots build furniture for us. Where AI creates flawless 3D designs and machines carve every curve flawlessly.

But that is not the point.

Hobbies like this exist outside of efficiency. They are not about quickness or precision. They are about the feeling. The challenge. The small wins along the way. The quiet satisfaction of sanding a piece until it is smooth, of fitting two imperfect boards together and realizing they still hold.

And woodworking is not the only place I have felt that.

When I ran my first half Ironman, the last mile of the run was neither efficient nor particularly pretty. My form was falling apart, my pace was nowhere near what I had trained for, and I am pretty sure I was mumbling encouragement to myself like a slightly deranged motivational speaker. But crossing that finish line? That rush of emotion was not about performance metrics. It was awe. It was the kind of pride that cannot be automated, as it stems from effort rather than output.

That is the beauty of human-made, human-earned things. The joy of holding something physical that did not exist until you created it. The pride of completing something difficult, even when it did not go according to plan. The connection that happens when you share that imperfect, deeply personal achievement with someone else.

Just like the joy of writing a blog post that resonates. Or recording a podcast episode that someone tells you helped them feel less alone.
Not everything needs to scale.

Some things just need to matter.

SIDEBAR: Questions AI Will Never Answer

- What does it mean to truly love someone?
- Why did that song make me cry?
- Am I living in alignment with my values?
- Should I forgive them?
- What kind of parent do I want to be?

You can ask AI these questions. It will answer. But it will not know. That knowing? That is yours alone.

And here is the thing: if we want more of those moments, we need to notice them when they appear. That is where emotional awareness comes in. Because awe, connection, and meaning do not show up on your calendar. They show up in flashes: a conversation, a finish line, a shared laugh, and you can miss them entirely if you are too busy chasing efficiency.

The Case for Emotional Awareness

Here is something I have learned over time: emotional awareness is a powerful skill. I was not always great at this. In fact, for a long time, my emotional process looked like: feel something → react immediately → regret it later.

It took time to realize that feelings are not instructions; they are signals. And if you learn to interpret that information before acting, you achieve better results, stronger relationships, and much less regret.

The good news? You can train for this, just like any other skill.

Step 1: Name It

The first step sounds simple but is surprisingly hard: name the emotion. Not just "good" or "bad." Get specific.

- Frustrated
- Anxious
- Lonely
- Overwhelmed
- Hopeful

By naming it, you make it concrete instead of letting it swirl around as a vague storm cloud in your head.

A standard tool that helps with this is called a *feelings wheel*. Imagine a circle that starts with broad categories like *angry, sad, happy, scared,* and then branches into more specific words as you move outward. For example, "sad" might expand into "lonely," "disappointed," or "helpless." The point is not to memorize every term; it is to notice that

132

your inner world has more nuance than just "fine" or "not fine."

Step 2: Locate It

Emotions do not just live in your mind; they show up in your body. Where are you feeling it? Tight chest? Restless legs? Knotted stomach? This might sound small, but noticing where you feel it makes you less likely to ignore it.

Step 3: Listen Before You Act

This is where the magic happens. Ask:

- What is this feeling trying to tell me?
- Is it alerting me to a real problem, or is it just a passing weather system in my brain?
- What would happen if I did not act on it right now?

That pause, even 10 seconds, creates space between you and the knee-jerk reaction.

Step 4: Choose With Intention

Once you have named, located, and listened to the feeling, decide how to respond, not how to react. That might mean having the conversation, taking a break, asking for help, or doing nothing at all.

Why This Matters in the Age of AI

Technology cannot do this for you. It can prompt reflection, offer calming techniques, even "simulate" empathy in a chat window. But it cannot feel what you feel, in your body, in your context, with your history. That is your edge as a human being.

When you build emotional awareness, you are not just becoming more "in touch"; you are improving your decision-making, deepening your relationships, and protecting your mental health.

And in a world obsessed with efficiency, being able to slow down long enough to ask, "What is really going on here?" is a power move.

The Human Edge

The most important things in life rarely arrive on schedule. They sneak up on us, in the middle of a conversation, in a sudden wave of gratitude, in the quiet moment before you drift off to sleep. They are not measurable in metrics or optimized by algorithms.

Awe, joy, grief, love; they are inconvenient. They interrupt your workflow. They do not care about your deadlines. And that is precisely why they matter.

Because these moments are evidence that you are alive. Proof that you are more than your output. Proof that your value does not come from efficiency, but from presence.

AI can replicate skills. It can scale speed. But it cannot feel what it is like to run into the ocean at sunrise, or watch your child take their first steps, or hear someone say, "I am glad you are here," and know they mean it.

That is the human edge. Not perfection. Not productivity. Presence. The willingness to feel, to connect, to be moved, and to let those moments change you.

Protect that. Practice that. And when the world tries to rush you past it, slow down. Because the stuff you cannot automate? That is the stuff that makes you, you.

From Self-Care to Shared Care

For years, the conversation around mental health has been dominated by *self-care*. Take a walk. Meditate. Journal. These practices matter as they ground us and restore us from the inside out.

But self-care has limits when practiced in isolation. It can keep you afloat, but it will not always pull you back to shore. For that, you need people.

I think of this as the shift from *self-care* to *shared care*. Self-care says: "I'll take a break so I can recharge." Shared care says: "I'll lean on you, and you can lean on me." One is a solo act; the other is a duet.

And science affirms it. Strong social ties don't just ease loneliness, they reduce the risk of depression, buffer stress, and even extend life expectancy [16, 17, 18]. No meditation app or productivity hack can replicate the protective power of being seen, heard, and supported.

In a culture that pushes hyper-individualism: your goals, your schedule, your "personal brand", it's worth remembering mental health is collective health. We heal in relationships. We recover in community. We thrive when care is something we share.

No algorithm can replicate shared care. It's one of the last, best things that makes us human, and it may be the very thing that saves us in an age of machines.

* * *

Clarity Check-in

Ask yourself:

- What part of your emotional life have you tried to ignore, suppress, or "optimize"?
- Why do you think you've been pushing it down? Fear, discomfort, or the pressure to stay "productive"?
- How do those unacknowledged emotions show up in your body? Tight chest, restless energy, headaches, fatigue?
- What would happen if you gave that emotion more space, not less?
- Which emotions feel hardest for you to sit with, and what might they be trying to tell you?

Experiment: Spend 10 minutes each day naming your emotions out loud or in a journal, without trying to fix or justify them. Use as many words as you can. If you struggle, borrow from a feelings list or wheel for inspiration. At the end of the week, notice: do your emotions feel heavier, lighter, or simply clearer when you give them room to exist?

Struggling to get past "good" or "bad"? No problem. In the Clarity Check-In section at the back, you will find a Feelings Vocabulary Starter, a quick list of emotions to help you name what you are really feeling.

* * *

9

The Future Is Not Fully Automated

Technology will evolve. You get to decide how you show up in it.

There was a time when "the future" was a hazy concept; flying cars, robotic butlers, maybe the occasional trip to Mars. It lived safely in science fiction, decades away.

Not anymore.

Now, the future is threaded through our everyday lives. Your watch tells you when to stand up. Your phone finishes your sentences. Your playlists know your mood before you do. You can chat with an AI that has read more books in an afternoon than you could in ten lifetimes.

It is exciting. It is unsettling. It is fast.

The future is not just happening to us. Every time we choose which tools to use, which tasks to automate, which decisions to outsource, we are quietly shaping the kind of humans we will be tomorrow.

A Digital Fork in the Road

There is a moment most of us do not notice until we are already living in it: the future is not coming. It is here.

It is in your pocket, buzzing with reminders. It is on your wrist, tracking your heart rate and nudging you to stand up. It is in your living room, answering trivia questions and controlling the thermostat. It is in your car, suggesting faster routes before you have even turned the key.

It is whispering while you type, drive, speak, or even breathe.

AI is not some distant sci-fi arrival. It is already here, quietly threading itself into our routines.

And every time we tap "yes" on a new update, let an app choose our playlist, or ask a voice assistant to tell us the news, we are doing more than saving time; we are training ourselves to let technology decide for us.

So here is the question no one else will ask for you:
What kind of human do you want to be in a machine-driven world?

Because make no mistake, you are making choices. Not just about efficiency. About identity.

A Tool, Not a Replacement

I love AI.

I use ChatGPT, Copilot, Grammarly, and other tools every single week. They help me brainstorm ideas I never would have landed on alone. They accelerate my outlines. They challenge my thinking. Sometimes they even rescue me from the dreaded blinking cursor.

But they are co-pilots, not captains.

That is not just semantics; it is survival.

You do not hand over your steering wheel to your hammer, no matter how well it drives nails. And yet it is easy to drift there: to let AI write the first draft, and the second, and the final, without checking if it still sounds like you.

That is why I try to think first, prompt second. Even if it is slower. I sketch my own outline. I jot my own messy ideas. I want my fingerprints on the work before a machine touches it.

And when my kids ask me a question, I resist the urge to call Alexa for backup. Sure, she could deliver a perfect answer. But I would rather give them my answer, even if it is flawed, because being human is not about flawless accuracy.

It is about showing up as your true self.

Everyday Automation Creep

Some handovers are obvious. Others are so subtle that you do not notice until you have outsourced half of your thinking.

Google Maps does not just tell you where to go; it decides which neighborhoods you will never drive through.

Spotify shapes your mood before you have had your coffee.

Instagram's feed subtly defines what you consider "normal" or "beautiful."

The "recommended for you" list becomes your default menu for movies, music, and even books.

Convenience is not bad. But when it is unexamined, it is invisible, and it is these invisible habits that shape us the most.

Your Digital Philosophy

If someone asked, "What is your digital philosophy?" could you answer?

Most of us cannot, because we have never thought about it. We adopt new tech without pause. New app? Download. New AI tool? Sign me up.

But without a philosophy, convenience becomes our only filter, and convenience does not care about your values.

Here is mine: embrace technology with curiosity and caution - and think through the implications before committing.

Before I add a tool, I ask myself:

- Will this make me more of the person I want to be?
- Am I outsourcing something I actually need to feel, learn, or wrestle with?
- Is this adding depth, or speed?

Self-awareness is your firewall.

Fast Is Not Always Better

There is a lot to love about AI. Let's not pretend otherwise. I love how quickly I can get a thoughtful response. I love how I can test a dozen ideas in seconds. I love that AI is becoming more accurate and nuanced.

But the quickness can be a trap.

Fast thinking is not always deep thinking, and efficiency does not always mean effectiveness. Just because we *can* automate something does not mean we *should*.

When a tool is fast, it is tempting to skip the pause; the space where you would normally check your gut, weigh the trade-offs, or see the bigger picture. That pause is also where ethics live.

Speed can make us skip fact-checking because "it sounded right." It can lead us to approve an AI-generated pitch without verifying whether it aligns with our actual values. It can make us rush to publish something clever that, in hindsight, was not true, kind, or even ours to share.

I once explored cloning my voice with AI for podcasting. I figured it would save time; no more editing, no more retakes. The software was impressive. It sounded eerily like me.

But it was not me.

It missed my cadence. My humor. My off-script moments. It lacked the imperfections that make human speech worth listening to. It saved time but lost the soul.

So I scrapped it. Because sometimes what is slower is actually richer.

AI can do a lot in milliseconds. But your judgment? That still needs minutes. Sometimes hours. And that is okay.

The Ethics of Automation

We often discuss what AI can do. We need to discuss more about what it should do.

AI is only as good as the data it is trained on, and that data comes from us: our histories, our biases, our blind spots. Which means those same biases can get baked right into the algorithms that now shape hiring decisions, loan approvals, medical diagnoses, and even the news headlines we see.

We have already seen what happens when no one is watching. A major tech company had to scrap its AI hiring tool after discovering it downgraded resumes from women because the training data was primarily male. Facial recognition systems have repeatedly shown higher error rates for people with darker skin tones, leading to false arrests. Social media algorithms have amplified misinformation and polarized communities, not because they are malicious, but because outrage and division keep people clicking.

These are not abstract risks. They are reminders that automation can magnify harm at a scale and speed humans simply cannot match. One flawed suggestion from an AI doctor, one biased investment recommendation, one inaccurate legal summary, all can spread faster and broader than any single human mistake.

That is why we need guardrails. Not just legal ones, but personal ones. Micro-ethics. The quiet decisions we make before hitting "generate," "send," or "share." Do I fact-check this before I post it? Should I mention to people that this email was AI-assisted? Do I understand where the data comes from and who benefits from me using it?

Because the truth is, an AI output is not neutral just because it is generated in 0.3 seconds. It reflects the fingerprints of whoever built it, trained it, and fed it. And if we do not pause to ask questions, those fingerprints become invisible, but their influence does not.

SIDEBAR: Questions, Questions, Questions

- Do I know where this AI's data comes from?
- Whose voices might be missing in the training set?
- Am I outsourcing a decision I actually need to feel, wrestle with, or take responsibility for?
- Who benefits most from me using this tool: me or the company behind it?

Ethics in automation is not about paranoia. It is about *awareness*. As our lives become more automated, we need to be more deliberate about what remains human.

Living by Your Values

So how do you future-proof your values in a world where everything feels up for grabs?

Start with presence.

When I write, I make sure the words are still mine. When I post, I keep my voice in it. When I record a podcast, I use my own voice: flaws, filler words, stumbles, and all.

When I parent, I model what it means to *be* here. To look someone in the eye instead of a screen. To discuss difficult topics instead of avoiding them.

I want my kids to grow up confident in who they are *before* the world tells them who they should be. That means no phones for a while. That means conversations at dinner. That means presence over performance.

Because values are not taught, they are shown.

SIDEBAR: Signs You are Letting Tech Define You

- You feel guilty when you are not being "productive."
- You have not made something with your hands in months.
- You cannot remember the last time you daydreamed.
- You trust AI more than your own gut.
- You check your phone first thing in the morning, and it is the last

thing you do at night.

These are not moral failings. They are signals. Check in with them.

Technology Can Enhance. But You Create Meaning.

Let's make this clear: AI does not create meaning. You do.

It can help write your resume, but it cannot tell you what job makes you come alive. It can suggest birthday gifts, but it does not know the inside jokes, the shared memories, the smile on your friend's face.

It can organize your to-do list, but it will not ask why you are doing any of it in the first place.

Technology is amazing. But without intention, it becomes noise. Without self-awareness, it becomes a distraction. Without values, it becomes control.

You need a personal philosophy not just for how you use tech, but for how you *live*.

Building Your Digital Compass

Here are a few principles I try to live by:

Slow down before you accelerate. Do not rush to integrate a new tool without first understanding its purpose. Not everything needs to be streamlined.

Start analog. Before I turn to tech, I try to reflect, journal, and brainstorm. I want my ideas to be *mine* before they get shaped by a machine.

Keep some spaces sacred. No screens at dinner. No phones before bed. These rituals matter more than any efficiency gain.

Let tech support your humanity, not replace it. If it helps you be more present, more focused, more kind? Great. Does it numb, distract, or dilute your life? Reconsider.

Teach values first. Tools second. Especially for our kids. They need to know who they are before they decide what to use.

Hope For the Human Future

There is so much to be excited about.

I genuinely believe AI will help us solve real human problems: in healthcare, education, access to services, and yes, mental health.

I hope it helps us detect cancer sooner. Connect isolated people to care. Reduce the weight of the tasks that drain us, so we have more time for what lights us up.

But the goal cannot just be *more*.

It has to be *better*. More connected. More alive. More honest.

My hope is that AI not only makes us more efficient but also reminds us of what only *we* can do.

Love. Forgive. Empathize. Create. Laugh. Sit in silence. Walk through grief. Choose joy.

That is not automation. That is being human.

* * *

Clarity Check-in

Ask yourself:

- If you had to write your personal philosophy for living well in an AI-enhanced world, what would it include?
- What do you want technology to amplify in your life? Clarity, creativity, connection, efficiency?
- What do you want to keep sacred? Rest, relationships, art, play?
- What would you never want to automate, even if you could?
- How will you know if you've crossed the line from helpful to harmful with your tools?

Experiment: Write a one-page "Human Charter" for yourself, a short list of guiding principles for how you want to live alongside AI. Keep it simple, no more than 5 lines. Revisit it every few months, and notice: what stays the same, and what shifts as both you and the tech evolve?

Not sure how to start? In the Clarity Check-In section at the back, you will find a Human Charter template, a simple way to sketch out the values and boundaries you want to keep in an AI-driven world.

* * *

10

Beyond Productivity, Toward Clarity

What if the goal is not to do more, but to see more clearly?

We have spent this book talking about focus, rest, boundaries, and the strange new role of AI in our lives. But if you have made it this far, you have probably realized this was never just a productivity book.

It is a book about being human while getting things done.

It is about clarity; choosing what matters, designing systems that protect your mind, and refusing to measure your worth by the rapidity of your output. It is about remembering that the real point of any productivity tool, digital or analog, is to give you more life, not more tasks.

Because the goal is not peak efficiency.

It is alignment. Presence. Meaning.

The Productivity Trap (Revisited)

We live in a culture where "busy" is worn like a badge, calendars are performance art, and rest feels suspicious. Burnout is treated like a rite of passage.

But productivity without clarity is just motion without direction. You can optimize every minute and still miss the moments that matter.

Clarity is the filter that says: This matters. That does not.
   It gives weight to your yes and power to your no.
   It allows you to stop when enough is enough, without guilt.

What Real Clarity Looks Like

Clarity does not always feel clean. It is not always about having a color-coded plan.

Clarity is saying no to a great opportunity because your values say otherwise. Clarity is closing the laptop even when the task is not done, because your kid asked you to play. Clarity is working on something deeply unsexy because it moves the needle in your soul, not just your metrics.

When you have clarity, you start building your life from the inside out, not from obligation, not from pressure but from alignment.

And that is where real productivity lives.

A New Set of Metrics

Here is a radical idea. What if we started measuring productivity by:

- How often we feel aligned with our values.
- How present we are with our loved ones.
- How well we treat our mental health.
- How free we feel in our own minds.

These are hard to measure. You cannot put them on a spreadsheet. But they matter more than your inbox zero streak.

The truth is that most people do not need more output. They need more *space*. More breathing room. More room to feel, to reflect, to redirect.

Clarity creates that space.

How to Keep Clarity in a Machine World

Clarity is not a one-time decision; it is a daily practice. The world will keep getting faster, tools will keep getting smarter, and your attention will keep getting pulled. That means clarity has to be protected, not just discovered.

Here is how I try to keep mine:

Audit your inputs. What you read, watch, and scroll shapes your thinking. If it is noise, cut it.

Protect white space. Leave gaps in your calendar on purpose. Clarity does not grow in overcrowded schedules.

Check your "why" before your "how". Tools are only as useful as the purpose they serve.

Set human-first defaults. Whether it is dinner without screens or keeping some tasks fully analog, decide where you want tech to stop.

Revisit your priorities. At least once a season, ask: "Does this still matter to me?"

These habits are not about resisting change; they are about steering it.

So, Where Do You Go From Here?

You do not need a 20-step morning routine or 15 productivity apps. You do not need to outsource your brain to AI or optimize every minute of your life.

You need to pause. Reflect. Ask better questions.

- What makes me feel most like myself?
- What is draining me that does not need to?
- What am I proud of that no algorithm could measure?
- Where do I want to focus my attention *on purpose*?

Because clarity is not the end goal, it is the beginning of everything that matters.

So take what serves you from this book. Leave what does not. And build your life like a human, not a machine.

You do not have to be perfect.

We won't out-hustle machines. But we can out-care them. Mental health isn't just self-care, it's shared care. And that's what keeps us fully human.

* * *

Final Clarity Check-in

Ask yourself:

- What would change in your life if you measured success by clarity, not productivity?
- When do you feel most clear-headed, and what conditions make that possible?
- How often do you chase output at the expense of peace of mind?
- If someone asked you what mattered most this week, would your calendar match your answer?
- What small shift could bring your daily actions closer to your deeper priorities?

Experiment: Each morning, write down the one thing that would make you feel clear-headed by bedtime, then make it your priority. At the end of the week, review your list. Did your "clarity priorities" look different from your usual productivity goals? What patterns or surprises do you notice?

* * *

Thank you for walking through this journey. For showing up. For staying human.

Epilogue

Still Human After All

Clarity is not the end of the journey; it is how you keep walking without losing yourself.

You were never meant to be perfect. You were meant to be real.

The world did not end when AI became more intelligent. But something else started happening. Slowly, subtly, invisibly. We began outsourcing not just our tasks, but our thinking. Our creativity. Our humanity.

And this book? It has been my pushback. My reminder: to you and to myself, that we do not have to surrender what makes us human just because the tools around us are evolving. We can still think deeply. We can still feel fully. We can still choose meaning over metrics.

You Are Not a Machine

You do not have to optimize every second of your day. You do not have to respond to every notification. You do not have to earn your rest.

The truth is, you were never meant to be "on" all the time. You were designed for rhythm, not relentlessness. For progress, not perfection.

You were meant to pause. To reflect. To stumble, to try again. To laugh at the mess.

And no matter what the algorithms say, no update will ever replace that.

Where We Go From Here

If you take one thing away from this book, let it be this:
   You do not need to be superhuman to succeed in a machine world. You just need to stay human.

And that starts with awareness. Knowing your limits. Knowing your values. Knowing the space between the constant reminders.

Let that awareness guide you. Let it shape the way you work, connect, rest, and create.

Because the future is not fully automated. And thank goodness for that.

You are still here. Still thinking. Still feeling.

Still human.

Clarity Check-ins

For easy access, here are all the prompts from each chapter in one place, including a few new ones, so you can revisit them whenever you need a reset.

These are not here to be "completed." They are invitations to pause, to explore, and to reconnect with the human behind the hustle.

Experiment: Pick one check-in question from this book each week for the next 3 months and actually act on your answer.

* * *

What kind of clarity are you truly seeking?
In a world full of tools, checklists, and algorithms telling you how to live and work "better," what does clarity actually mean to *you*? What is worth focusing on, and what is just noise?

* * *

How is your relationship with technology shaping your relationship with yourself?
Think about your daily routines, your creative habits, even your inner

dialogue. Where is tech helping you become more *you*, and where is it quietly hijacking your attention or values?

* * *

What parts of your emotional life have you tried to optimize, suppress, or ignore?
When do you feel most like a human, not a machine? How can you give your emotions more space, rather than trying to manage them like a to-do list?

* * *

Where are you confusing productivity with worth?
How much of your identity is tied to output? What would it look like to be just as valuable, even if you accomplished nothing "visible" today?

* * *

What kind of future do you want to help build, for yourself, your family, your work, your world?
We cannot stop change. But we can shape our role in it. What does it mean for *you* to stay fully human in a machine-driven world?

* * *

From Chapter 1: Your Brain Was Not Built for This

Take a moment to scan your current digital life:

- Which tools or devices genuinely support your well-being and productivity?
- Which ones leave you feeling more scattered, stressed, or drained: even if they are marketed to "help"?
- When you think about your typical day, which alerts or notifications feel urgent but rarely matter?
- If your brain could write you a one-sentence message today, what would it say?
- How does your body feel during heavy screen use? Tense shoulders, racing heart, tired eyes? What might that be telling you?
- If you could protect one part of your day from the digital world, what would it be? Mornings, meals, evenings, or rest?

Experiment: Pick one "maybe helpful" app, device, or alert and turn it off for 24 hours. Pay attention not just to what is missing, but to what shows up in its place: a calmer morning, fewer distractions, a surprising burst of focus.

Write it down. Don't overthink it. Just listen.

* * *

From Chapter 2: Productivity is Not Peace
Think about your current approach to productivity:

- What does a "successful day" look like for you right now?
- Does that definition include your well-being or just your output?
- How do you usually feel at the *end* of a productive day? Satisfied,

drained, restless, calm?
- What trade-offs are you making to keep up? (sleep, relationships, creativity, downtime)
- What would it look like to measure your day by energy, peace, or purpose instead of just completed tasks?
- If you zoomed out to look at your week or month, what would you actually want to see more of? (connection, rest, focus, joy, progress)

Experiment: Write your own definition of productivity, one that actually supports your mental health. Keep it short, no more than two sentences. Then, for the next week, try checking your day against *that* definition instead of your to-do list.

* * *

From Chapter 3: When Tech Starts to Think for You
  Think about your own use of AI or productivity technology:

- What kinds of thinking or tasks do you most often delegate to tools?
- Has AI ever helped you uncover something you could not reach on your own, or has it ever made your work feel *less like you*?
- Which parts of your work or creativity feel too important to out-source?
- Do you trust the answers your tools give you, or do you always double-check?
- How can you use AI as a co-pilot without giving it the steering wheel?
- If your favorite tool disappeared tomorrow, what skills or practices would you still want to hold onto?

Experiment: Write a short note to your future self: What do you want to remember about staying human in a world that is getting smarter? Save it somewhere: in your journal, in your note's app, or even email it to yourself. Revisit it in six months and see if you have kept the balance you hoped for.

* * *

From Chapter 4: Focus is Your Superpower
Think about a moment recently when you felt truly focused:

- Where were you? What were you doing?
- What distractions were *not* there and how can you recreate that condition more often?
- How did your body feel in that state (calm, energized, absorbed)?
- What usually pulls you away from focus? Notifications, people, your own thoughts?
- What part of your day feels most scattered, and what small boundary could help protect it?
- If you could design a "focus ritual", one habit or environment cue that tells your brain it is time to lock in, what would it be?

Experiment: Write down one "focus shift" you want to try this week. Keep it small. Keep it honest. At the end of the week, notice: did it help, hinder, or just need tweaking?

* * *

From Chapter 5: Rest Like You Mean It
Think about your current rest habits:

- When was the last time you truly felt restored: mentally, emotionally, or physically?
- What stops you from resting when you need to? Is it guilt, pressure, or fear of falling behind?
- How do you usually try to rest and does it actually recharge you, or just numb you?
- What signals does your body give you when it is running low on energy?
- What small boundary or ritual could you add this week to protect your energy?

Write a "Rest Permission Slip" for yourself. Keep it visible. Use it when guilt shows up.

Experiment: Schedule a 20-minute *active rest* break daily: walk, stretch, quiet teatime, and treat it like a meeting you cannot cancel. At the end of the week, reflect: did those breaks shift your energy, mood, or focus?

Energy Audit (Self-Reflection)

Your brain and body don't run on a flat battery. Energy rises and falls throughout the day: physically, mentally, emotionally, and socially. Doing a quick audit can help you spot patterns so you know when to push forward and when to protect your rest.

Step 1: Map Your Energy

Rate your energy levels at different times of the day on a scale from **1 (very low)** to **5 (very high).**

Step 2: Reflect

Ask yourself:

- When do I feel most energized?
- What tasks or activities drain my energy the most?
- What activities, habits, or people help me recharge?

Step 3: Notice Your Patterns

Energy Drains (watch out for these):

- Back-to-back commitments with no breaks
- Unclear expectations
- Excessive screen time
- Poor sleep habits

Energy Gains (build more of these in):

- Taking short breaks and moving around
- Drinking water and eating nourishing snacks
- Engaging in positive conversations
- Setting clear goals and priorities

* * *

From Chapter 6: The New Rules of Getting Things Done

Getting things done is not about checking more boxes, it is about

checking in with yourself.

- Let your tools serve you, not run you. Let your calendar support your mind, not hijack your time. And remember: you are still the point of the process.
- Ask yourself:
- What is one tool, routine, or mindset you have been using that adds more stress than clarity?
- When you open your calendar or task list, do you feel supported or overwhelmed?
- Can you simplify it? Swap it out? Let it go?
- If you rebuilt your system from scratch, what would you *actually* keep?
- Now, write down one "new rule" you want to live by. Not to do more, but to do what matters.

Experiment: For one week, limit yourself to 3 meaningful tasks per day. At the end of each day, track how you feel: more in control, less? More energized, less? By the end of the week, notice whether doing *less* actually helped you achieve *more of what matters*.

Your New Rules Starter Template

Take a moment to sketch out a few "new rules" for yourself. Keep them simple, personal, and doable. These are not aspirational slogans. These should guide how you want to approach your work, your time, and your energy.

Examples:

I will measure my day by energy, not just output.
I will protect my mornings from email.

I will end the workday with one task unfinished if it means I can rest.

Your Turn:

New Rule #1: _____

_____

New Rule #2: _____

_____

New Rule #3: _____

_____

(Optional) **New Rule #4:** _____

_____

(Optional) **New Rule #5:** _____

_____

* * *

From Chapter 7: Imperfection is the Point – Why Your Flaws Are Your Superpower

Ask yourself:

- What is one area of your life where you are holding yourself to an unrealistic standard?
- Where does perfectionism show up most often? Work, parenting, relationships, or even hobbies.
- What is the cost of chasing perfection? (lost energy, delayed progress, missed joy)
- Whose approval are you actually trying to earn, and does it matter as much as you think?
- What would happen if you let go of that perfection, just a little, and gave yourself permission to show up as you are?

Experiment: This week, share or publish one piece of work "as is". No extra polishing, no endless tweaks. Then, notice how people respond compared to your usual process. More importantly, how do *you* feel? Lighter, anxious, relieved, or maybe even proud? Capture that response and remind yourself: imperfect still counts.

* * *

From Chapter 8: Emotions, Meaning, and the Stuff You Cannot Automate
  Ask yourself:

- What part of your emotional life have you tried to ignore, suppress, or "optimize"?
- Why do you think you have been pushing it down? Fear, discomfort, or the pressure to stay "productive"?
- How do those unacknowledged emotions show up in your body? Tight chest, restless energy, headaches, fatigue?
- What would happen if you gave that emotion more space, not less?
- Which emotions feel hardest for you to sit with and what might they be trying to tell you?

Experiment: Spend 10 minutes each day naming your emotions out loud or in a journal, without trying to fix or justify them. Use as many words as you can. If you struggle, borrow from a feelings list or wheel for inspiration. At the end of the week, notice: do your emotions feel heavier, lighter, or simply clearer when you give them room to exist?

Feelings Vocabulary Starter
  When you are naming emotions, it helps to go beyond the basics. Here

are some words you can borrow if you get stuck:

Energized / Upbeat

hopeful

excited

curious

confident

Calm / Content

peaceful

relaxed

grateful

connected

Tense / Uneasy

anxious

overwhelmed

restless

pressured

Low / Heavy

sad

lonely

disappointed

drained

Irritated / Activated

frustrated

impatient

angry

defensive

* * *

From Chapter 9: The Future is Not Fully Automated

Ask yourself:

- If you had to write your personal philosophy for living well in an AI-enhanced world, what would it include?
- What do you want technology to amplify in your life? Clarity, creativity, connection, efficiency?
- What do you want to keep sacred? Rest, relationships, art, play?
- What would you never want to automate, even if you could?
- How will you know if you have crossed the line from helpful to harmful with your tools?

Experiment: Write a one-page "Human Charter" for yourself, a short list of guiding principles for how you want to live alongside AI. Keep it simple, no more than 5 lines. Revisit it every few months, and notice: what stays the same, and what shifts as both you and the tech evolve?

Your Human Charter Template

Write down a few guiding principles for how you want to live alongside AI and other emerging tools. Keep them short, personal, and human-centered. These are not rules for doing more, they are reminders of what you want to protect, preserve, and prioritize.

Examples:

- *I will use AI to support my creativity, not replace it.*
- *I will keep rest, play, and relationships sacred. They are not for automation.*
- *I will choose tools that reduce stress, not add to it.*
- *I will measure progress by meaning, not just speed.*

Your Turn:

Human Charter Principle #1: _____

_____

Human Charter Principle #2: _____

_____

Human Charter Principle #3: _____

_____

(Optional) **Human Charter Principle #4:** _____

_____

(Optional) **Human Charter Principle #5:** _____

_____

* * *

From Chapter 10: Beyond Productivity, Toward Clarity
Ask yourself:

- What would change in your life if you measured success by clarity, not productivity?
- When do you feel most clear-headed and what conditions make that possible?
- How often do you chase output at the expense of peace of mind?
- If someone asked you what mattered most this week, would your calendar match your answer?
- What small shift could bring your daily actions closer to your deeper priorities?

Experiment: Each morning, write down the one thing that would make you feel clear-headed by bedtime, then make it your priority. At the end of the week, review your list. Did your "clarity priorities" look different from your usual productivity goals? What patterns or surprises do you notice?

Your Clarity Manifesto

This is your chance to distill everything you have reflected on into a short, guiding statement. Think of it as your compass for living with more presence, balance, and humanity in an AI-driven world.

Examples:

- *I choose clarity over busyness.*
- *I measure my life by energy, connection, and meaning; not just output.*
- *I will rest without guilt, focus without apology, and allow imperfection to be part of my strength.*
- *I will use technology as a tool, not a taskmaster.*
- *I will protect what makes me human: empathy, creativity, and emotion.*

Your Turn:

Clarity Manifesto Statement #1: _____
_____

Clarity Manifesto Statement #2: _____
_____

Clarity Manifesto Statement #3: _____
_____

(Optional) **Clarity Manifesto Statement #4:** _____
_____

(Optional) **Clarity Manifesto Statement #5:** _____
_____

Keep your mind strong, your coffee stronger, and your clarity intact.

Resources for the Real World

Practical tools, trusted support, and recommendations to help you navigate the topics explored in this book.

Mental Health Support

Emergency and Crisis Lines:

- 988 Suicide & Crisis Lifeline (U.S.): Dial 988 anytime, 24/7
- Crisis Text Line: Text HOME to 741741 (U.S.)
- SAMHSA Helpline: 1-800-662-HELP (4357) – for treatment referrals and information

* * *

Online Therapy Platforms:

- Calmerry (affiliate partner)
- BetterHelp
- Talkspace
- Online-Therapy.com (affiliate partner)

* * *

Directories and Resources:

- NAMI – National Alliance on Mental Illness
- Psychology Today – Find a therapist near you
- Mental Health America – Screening tools and advocacy

* * *

Tools for Clarity and Focus

Daily Planning & Productivity:

- Sunsama – Daily planner integrating tasks, calendar, and focus time (affiliate partner)
- Todoist – Task management tool (affiliate partner)
- Notion – All-in-one workspace for notes, tasks, wikis
- Trello – Visual project and task boards

* * *

Focus and Attention:

- Forest – Gamified focus app that plants real trees

- Freedom – Distraction blocker for websites and apps
- Pomofocus – Simple Pomodoro timer for deep work

* * *

Well-Being & Self-Care:

- Headspace – Guided meditations and mindfulness
- Calm – Sleep, meditation, and relaxation tools
- Insight Timer – Free meditations from global teachers

* * *

Helpful AI Tools

For Writing and Creativity:

- ChatGPT (OpenAI) – Brainstorming, outlining, first-draft generation
- Grammarly – Writing improvement and grammar suggestions

* * *

For Productivity and Automation:

- Microsoft Copilot – Embedded AI within Microsoft 365 tools
- Zapier – Automate tasks between apps
- Otter.ai – Meeting transcription and summary tool

* * *

Stay Connected

Visit www.thementallens.com for:

- Weekly blog posts on mental clarity and focus
- Podcast episodes from **Through the Mental Lens**
- Subscribe to the newsletter for *Free* toolkits and email updates

* * *

You are not alone in this journey. These resources are here to support your mind, not just your output.

Research Notes & Sources

Below are the studies, reports, and research findings referenced throughout *Still Human*.

1. **Van Vugt, M., & Li, N. P. (2018).** *The Evolutionary Mismatch Hypothesis: Implications for Psychological Science.* Current Directions in Psychological Science.

2. **Cowan, N. (2010).** "The Magical Mystery Four: How is Working Memory Capacity Limited, and Why?" *Current Directions in Psychological Science*, 19(1), 51–57.

3. **Rubinstein, J. S., Meyer, D. E., & Evans, J. E. (2001).** "Executive Control of Cognitive Processes in Task Switching." Journal of Experimental Psychology: Human Perception and Performance, 27(4), 763–797. Research summarized by the American Psychological Association indicates that task switching can reduce productivity by up to 40% and increase error rates.

4. **DeFilippis, E., Impink, S. M., Singell, M., Polzer, J. T., & Sadun, R. (2020).** *Collaborating During Coronavirus: The Impact of COVID-19 on the Nature of Work.* Harvard Business School Working Paper, No. 20-138.

5. **Gallup. (2021).** Employee Burnout: Causes and Cures. Gallup Workplace Report. Research indicates that 76% of employees experience burnout at least sometimes, and that lack of clarity

and unmanageable expectations are stronger predictors than workload.

6. **Google (2025).** AI Works 2025: Workplace Generative AI Report. Survey data indicates that 34% of workers use generative AI for work, and 75% of those users access it multiple times per week. Source: https://publicpolicy.google/resources/ai_works_2025_en.pdf

7. **Dell'Acqua, F., et al. (2023).** "Generative AI and the Decline of Independent Reasoning." *Massachusetts Institute of Technology*. MIT Study on AI and Critical Thinking. Research indicates that students who use AI tools over time demonstrate weaker critical thinking skills compared to their peers who do not use AI.

8. **Lee, M., et al. (2024).** "The Effects of Generative AI on Creative Output." *Stanford University*. Stanford Study on AI and Creativity in 2024 showing that participants relying on AI for first drafts experienced a measurable decrease in originality over time.

9. **Bhatia, S., et al. (2023).** "Perceived vs. Actual Quality in Human–AI Collaboration." *Nature Human Behaviour*, 7(8), 1124–1136. Nature Human Behaviour Study. Findings show that people using generative AI in collaborative work rated their ideas more favorably than objective assessments indicated, suggesting inflated confidence alongside reduced quality.

10. **Deloitte (2023).** *2023 Global Mobile Consumer Survey.* Found that the average adult checks their phone 144 times per day, illustrating the cumulative toll of frequent device interaction on focus and mental presence.

11. **Ophir, E., Nass, C., & Wagner, A.D. (2009).** "Cognitive Control in Media Multitaskers." *Proceedings of the National Academy of Sciences*, 106(37), 15583–15587. Stanford Multitasking Research. A Study demonstrating that chronic multitaskers perform worse on memory and attention tasks than those who focus on one task

at a time.

12. **X. Lin, W.-S., & Chao, T.-Y. (2025).** From Efficiency to Overload: Examining the Impact of ICT Multitasking on Work Performance and Executive Functions. AMCIS 2025 Proceedings. Research suggests that multitasking initially boosts performance but leads to cognitive and emotional costs, such as working memory deficits and attention residue.

13. **Y. Wannagat, W. (2024).** Media multitasking: Performance differences between younger and older adults. Media multitasking impairs cognitive performance across age groups.

14. **Raichle, M. E., & Gusnard, D. A. (2002).** "Appraising the Brain's Energy Budget." *Proceedings of the National Academy of Sciences*, 99(16), 10237–10239. It has been found that although the brain represents only ~2% of body weight, it consumes about 20% of the body's resting metabolic energy.

15. **Schulte, B. (2014).** Overwhelmed: Work, Love, and Play When No One Has the Time. Sarah Crichton Books. Schulte describes "time confetti" as the minor, constant interruptions that fragment our days into unproductive scraps of attention.

16. **Waldinger, R. J., & Schulz, M. S. (2010).** What makes for a good life? Lessons from the longest study on happiness. *Harvard Study of Adult Development*, Harvard University.

17. **Holt-Lunstad, J., Smith, T. B., & Layton, J. B. (2010).** Social relationships and mortality risk: A meta-analytic review. *PLoS Medicine*, 7(7), e1000316.

18. **Teo, A. R., Choi, H., & Valenstein, M. (2013).** Social relationships and depression: Ten-year follow-up from a nationally representative study. *PLoS ONE*, 8(4), e62396.

19. **Mental Health America. (2023).** *Stress in America: How Americans cope and where they turn for support.* Mental Health America.

Acknowledgments

I still cannot believe I wrote a book. Honestly, after tinkering with a cookbook since 2020, this feels like a bona fide miracle. And it is a miracle made possible by the people who reminded me, day after day, what it means to be fully, messily, beautifully human.

To my wife, Laura. You are my love, my compass, my calm, and my accounta-bila-buddy. *Yes we have a made-up word for holding each other accountable, because plain English just did not cut it.* You always make me laugh and remind me not to take myself too seriously. You were both my first fan and my toughest editor. I think Google Docs is still recovering from all your comments. I would not be who I am today without your love, support, and, most of all, your patience. Your compassion reminds me of what staying human really means.

To my boys, you are the "why" behind everything I do. I hope this book shows you that dreams are worth chasing, that mistakes are worth making, and that your voice matters in a noisy world. My most profound hope is to leave a legacy you can be proud of. I love you more than words on a page can hold.

To my peer reviewers Christy Long, Ivan Pais, Melissa Garabedian, Jennifer Rist, and Nick Bramel: thank you for your honesty, your sharp eyes, and your willingness to tell me when something was not working; and for pointing me toward better ways of saying what I meant. Your feedback shaped this book in ways I could not have done alone.

To the readers of *The Mental Lens* blog and listeners of *Through the Mental Lens*, your messages, comments, and quiet encouragement have been fuel for this work. Every "this helped me" note you have sent has been a reminder to keep going. This book is for you and because of you.

To those I have worked with, past and present, who have modeled both hustle and humanity, thank you for showing me that ambition and rest can coexist. And a special thanks to my friend EE for inspiring me to set my own boundaries, *and stick to them.*

And finally, to anyone who has ever felt overwhelmed, burned out, anxious, or not enough, but showed up anyway; this is for you. We are all just figuring it out, and that is more than enough. If this book does nothing else, I hope it reminds you that your humanity is your greatest strength.

About the Author

Chris Cage is a product manager, writer, endurance athlete, and the voice behind *The Mental Lens* blog and the podcast *Through the Mental Lens*. After years in high-pressure environments that prized efficiency over empathy, Chris set out on a journey toward clarity and connection.

Blending experience from healthcare tech, Ironman training, parenting, and his own mental health struggles, Chris shares tools and stories that encourage readers to embrace imperfection, protect their energy, and thrive as whole humans, even in an AI-driven world.

When he is not writing or recording episodes, you will likely find him carting his kids around to endless activities, building furniture in his garage, or logging long training runs that make him question why he keeps signing up for races requiring swim caps and energy gels.

Listen to *Through the Mental Lens* on Spotify, Apple Podcasts, or wherever you get your podcasts.

You can connect with me on:

🌐 https://www.thementallens.com

📘 https://www.facebook.com/thementallens

🔗 https://www.instagram.com/thementallens

🔗 https://throughthementallens.buzzsprout.com

Subscribe to my newsletter:

✉ https://www.thementallens.com